Wireless Mobile Phone Access to the Internet

INNOVATIVE TECHNOLOGY SERIES
INFORMATION SYSTEMS AND NETWORKS

Wireless Mobile Phone Access to the Internet

edited by
Thomas Noel

KOGAN
PAGE
SCIENCE

London and Sterling, VA

First published in Great Britain and the United States in 2003 by Kogan Page Science, an imprint of Kogan Page Limited

120 Pentonville Road
London N1 9JN
UK
www.koganpagescience.com

22883 Quicksilver Drive
Sterling VA 20166-2012
USA

© Kogan Page Limited, 2003

The right of Thomas Noel to be identified as the editor of this work has been asserted by him in accordance with the Copyright, Designs and Patents Act 1988.

ISBN 1 9039 9632 5

British Library Cataloguing-in-Publication Data

A CIP record for this book is available from the British Library.

Library of Congress Cataloging-in-Publication Data

Wireless mobile phone access to the Internet / edited by Thomas Noel.
      p. cm. -- (Innovative technology series: information systems and networks)
Includes bibliographical references and index.
  ISBN 1-903996-32-5
  1. Wireless Internet. 2. Cellular telephone systems. I. Noel,
Thomas, 1970- II. Series.
  TK5103.4885 .W5725 2003
  004.67'8--dc21
                                        2002151385

Typeset by Saxon Graphics Ltd, Derby
Printed and bound in Great Britain by Biddles Ltd, Guildford and King's Lynn
www.biddles.co.uk

Contents

Foreword

Wireless mobile access to the Internet will add a new dimension to the way we access information and communicate. Currently, several network protocols are being studied to maintain communication between a mobile host and its correspondents during mobile host movement. This book is devoted to the presentation of recent research on the deployment of the network protocols and services for mobile hosts and wireless communication on the Internet.

Today many people are connected to the Internet through mobile nodes like the palmtop or laptop. They need protocols to support new applications (like multimedia, voice or video over wireless). There are lots of new protocols and extensions to improve the IP connectivity of mobile nodes. These proposals aim to reduce the latency and the packet loss due to handoff between one point of attachment to another and to reduce the signaling overhead.

At the same time, a lot of wireless technologies have been developed: IEEE 802.11b, Bluetooth, HiperLAN/2, GPRS, UMTS. All of them have the same goal: offering wireless connectivity with minimum service disruption between mobile handovers. Of course, these handovers happen between access points located on the same sub-network. To support handovers between several sub-networks or distinct technologies a new protocol is needed. The IETF (Internet Engineer Task Force) has therefore defined the mobile IP protocol. Mobile IP is designed to manage mobile node movements between wireless IP networks. This protocol supports basic functionalities and it thus needs improvements to support optimized and secure communications.

The mobile world is divided into two parts: on one hand we have mobile nodes attached to several access points during mobiles' movement. These networks, called wireless networks, are interconnected by wired networks. On the other hand, we have ad-hoc networks which do not use any infrastructure to communicate. All nodes are mobiles and they cooperate in order to forward information between each other. This book presents these two worlds and introduces some research papers that propose extensions and optimizations to the existing protocols for mobility support.

Some of the papers introduce some concepts to improve networks with mobile stations. They propose to use some new concepts like security with AAA protocol, header compression and quality of service. They also introduce new functionality in core networks to discover new services.

Moreover, to develop a new protocol, we need some results to determine if it introduces new real possibilities, supports more users per access point, increases bandwidth, etc. Some simulations in this book describe optimizations for wireless access medium protocol. They also introduce results on ad-hoc network unicast routing protocol.

Finally, we can suppose that in the near future new mobiles will appear. They will support multiple wireless interfaces. For example, these ubiquitous terminals will use simultaneous IEEE 802.11b/a and Bluetooth technologies. Naturally, the increase of mobile nodes and number of interfaces per terminal will require a large addressing space, which is today unfortunately not available with the current Internet Protocol. Therefore, the new version of the Internet Protocol (IPv6) would be one of the next challenges for the wireless community.

Thomas Noel

Chapter 1

Distributed context transfer framework for mobility support

Hamid Mahmood Syed and Gary Kenward
Nortel Networks, Ontario, Canada

1. Introduction

There are a large number of IP access networks that support hosts that are not fixed – wireless Personal Area Networks (PANs), wireless LANs, satellite WANs and cellular WANs for example. The nature of this mobility is such that the communication path to a host may change frequently and rapidly.

In networks where the hosts are mobile, the forwarding path through the access network must often be redirected in order to deliver the host's IP traffic to the new point of access. The success of time sensitive services like VoIP telephony, video, etc. in a mobile environment depends heavily upon the minimization of the impact of this traffic redirection. In the process of establishing the new forwarding path, the nodes along the new path must be prepared to provide similar forwarding treatment to the IP packets.

The information required in support of a specific forwarding treatment provided to an IP flow is part of the context for that flow. This context is comprised of configuration and state information. To minimize the impact of a path change on an IP flow, the context must be replicated from the forwarding nodes along the existing path to the forwarding nodes along the new path.

The transfer of context information may be advantageous in minimizing the impact of host mobility on, for example, AAA, header compression, QoS, Policy, and possibly sub-IP protocols and services such as PPP. Context transfer at a minimum will be used to replicate the configuration information needed to establish the respective protocols and services. In addition, it will also provide the capability to replicate state information, allowing stateful protocols and services at the new node to be activated along the new path with less delay, and possibly less packet loss and less signaling overhead. There are few solutions available on the subject of context transfer [CNP] [CT-802.11] [CT-ICMP] (some of which are link layer technology-specific). The Seamoby WG at the IETF captured a problem statement [2] on how context transfer could be useful in minimizing service disruption.

This paper proposes a generic framework to replicate the context associated with the MN's traffic. The framework does not make any assumptions on the underlying sub-IP layer technologies. It identifies the various functional elements and the protocols required to exchange information between the functional entities, and capture the specific required characteristics for these protocols. Section 2 of the paper presents reference architecture for the mobile wireless access network. The various terms used in the paper are defined in Section 3. Section 4 captures details of the framework. Section 5 is targeted for providing example scenarios to explain the when and how of the communication between various functional entities enables a context transfer. A conclusion and some future work discussion follow in Sections 6 and 7. References to the in-progress IETF drafts on the subject are also provided at the end.

2. Wireless Access Network: reference architecture

Figure 1 shows the various elements of the mobile wireless access network. These entities only represent the functional elements and no mapping to physical boxes is assumed here. There can be a variety of ways to map these functional entities on to the physical network elements.

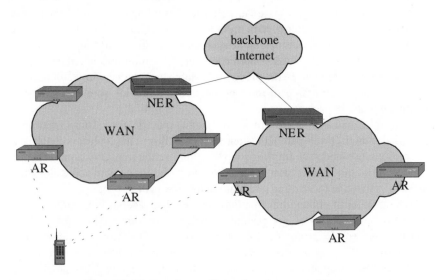

Figure 1. *Wireless Access Network: reference architecture*

2.1 Mobile Node (MN)

The Mobile Node represents an IP node capable of changing its point of attachment to the network. The MN can be either a mobile end-node or a mobile router serving an arbitrarily complex mobile network.

2.2 Access Router (AR)

The Access Router represents an IP router residing in an Access Network. An AR offers IP connectivity to MNs. An AR may include intelligence beyond the simple forwarding service offered by ordinary IP routers.

2.3 Network Edge Router (NER)

The Network Edge Router provides IP gateway functions that separate the Access Network from a third party network. It basically connects the Access Network to the global Internet.

2.4 Access Network (AN)

The Access Network represents an IP network that includes one or more ARs and NERs.

3. Terminology

There are a few terms that are useful in the discussion of context transfer framework. A brief definition of each term is included in this section.

3.1 Proactive context transfer

The context information required to completely support an IP micro-flow is replicated to the Access Router(s) that detect the presence of MN in its coverage area, in advance of the first packet arrival to one or any of these ARs. A *proactive context transfer* can be performed for a make-before-break or for a break-before-make packet forwarding scenario.

3.2 Reactive context transfer

The context information required to completely support an IP micro-flow is replicated to the Access Router at the instant when a packet from that micro-flow arrives at the new Access Router. A *reactive context transfer* can be performed for a make-before-break or for a break-before-make packet forwarding scenarios.

3.3 Configuration context information

The component of the context information that remains unchanged throughout the life of the session. Examples of such information are the bandwidth requirements of the application, differentiated services code point, some authorization information that is required only once for the MN.

3.4 State context information

The component of the context information which requires an update with every new packet arrival to the Access Network. Examples of such information include the accounting and header compression information at the Access Router.

3.5 Context group

A context group represents a logical grouping of Access Routers. It is comprised of the ARs currently supporting a given MN's micro-flow(s), and those ARs that have been identified as candidates for the imminent handoff of the micro-flow(s). As described, the latter ARs must all have a replicate of the necessary context before the MN's micro-flow(s) can be supported properly.

A member of a context group may never be required to forward traffic from the MN. The members of a context group may or may not be located in the same subnet or even the same administrative domain. However, there must be a communications path between all members of the group, and these paths must support the context transfer protocols.

4. A distributed framework for context transfer

This section provides the details of a framework to replicate the context associated with a MN's traffic flows to one or more receiving ARs. The proposed framework identifies the functional entities participating in the context transfer and describes the role of each functional entity. The context transfer framework includes definition of one key network event and two protocols.

4.1 Functional elements of the framework

The major functional elements of the context transfer framework are the context transfer agent (CTA), and the membership collection and distribution function (MCDF). Figure 2 shows a simple configuration involving these two functional elements and the protocols required.

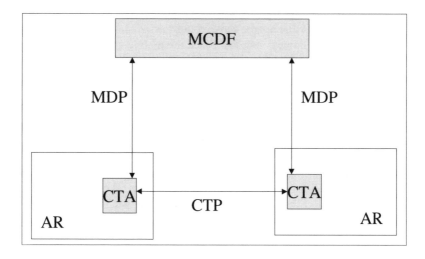

Figure 2. *Context transfer framework: functional entities and protocols*

4.1.1 Context Transfer Agent

The *context transfer agent (CTA)* is the functional entity that actively participates in the process of context transfer. Residing at the peers of the context transfer, the CTA is responsible for collection of all of the information required to support the active microflow associated with the MN to create the overall context that needs to be transferred. The CTA stores this context information in a form required by the context transfer protocol to carry between the peers of the context transfer.

The CTA also communicates with the membership collection and distribution function (MCDF) assisting it in context group management. The CTA exchanges the membership information with MCDF through a standardized membership distribution protocol (MDP).

The CTA residing at the receiver of the context information distributes the received information to the relevant entities for processing. The first operation to be invoked at the receiver on reception of context is the admission control. The CTA may help in intelligent handover decision-making by relaying the outcome of the admission control decision to the handover decision point HODP (cf Section 5.1).

4.1.2 Membership Collection and Distribution Function (MCDF)

The *membership collection and distribution function (MCDF)* is a functional entity that administers the formation of context groups. It creates and maintains one context group per active MN. The MCDF also collects the identification of the ARs that are supporting the active microflows of the MN and stores them as potential sources of the context for the MN.

For both proactive and reactive modes, the candidate AR needs to know the identification of the AR that has the active context associated with the MN's traffic i.e., the source of the context transfer. Since the mobile arrival to an AR in the access network is not predictable, this identification of the sources and candidates has to be performed on a real-time basis. When an AR receives a notification of the MN's arrival in its communication area (thus becomes a context group candidate, CGC), it communicates with the MCDF (through MDP) to get the information of the source of the context (CTS) for the MN. On the other hand, an AT that discovers himself as the source of the context by establishing active context associated with the MN's traffic, registers with the MCDF as the source for MN's context.

The MCDF function may be mapped to a single physical entity in the network or may be distributed throughout the network. If the MCDF is distributed, a standardized protocol must be defined to support the distribution and coordination of the membership information.

4.2 Protocols/events

The context transfer framework includes the definition of two key network events and two protocols.

4.2.1 Mobile Arrival/Departure Event (MADE)

The *mobile arrival/departure event (MADE)* is a notification delivered to an AR when the MN enters or leaves the AR's communication area. The requirements for such an event definition have been captured in [3]. The arrival or departure event enables the AR to join or leave a context group. The AR reacts to the arrival event by initiating a communication with the MCDF (as the context group candidate) for the identification of the source of the context. Similarly, a departure event initiates the removal of the AR from the context group associated with the MN.

4.2.2 HandOver Event (HOE)

The *handover event (HOE)* is a layer 3 notification that enables the HODP (cf Section 5.1) to initiate a handover from the source AR to the candidate AR. The reception of HOE also indicates that all the phases of the context transfer must be completed, ideally, before the first packet arrival at the candidate AR.

4.2.3 Membership Distribution Protocol (MDP)

The *membership distribution protocol (MDP)* enables the MCDF to effectively manage the context groups. It exchanges the context group membership with the CTA.

The MDP carries the following information:

- identification of the mobile node;
- context group information;

- identification of the AR (or ARs) that is to act as the source of the context transfer (CTS).

4.2.4 Context Transfer Protocol (CTP)

The *context transfer protocol (CTP)* provides mechanisms to transfer context information for a MN's traffic from the source (CTS) to the receiver, the CGC (context group candidate). The requirements for such a protocol are captured in [SYE].

5. Inter-working with the mobility framework

The goal of reducing the effect of mobility on the service level received by the MN through context transfer cannot be achieved without providing a close inter-working between the context transfer process and any mechanism that is responsible for the actual traffic re-direction from one AR to the other. For this reason, discussion of an interface between the context transfer and the mobility framework will help clarify the interaction between the two frameworks. The context transfer framework discusses the roles of one functional entity in the mobility framework and an interface with the mobility framework. This entity in the mobility framework may have different names for different mobility mechanisms. The context transfer framework, however, is a generic framework that should inter-work with any mobility mechanism.

5.1 HandOver Decision Point (HODP)

The *handover decision point (HODP)* represents a functional capability that is provided by the mobility framework. For a given mobility solution, it will likely have a different name and may be effected by a collection of network elements. The HODP is the entity that determines which of the ARs would be the best candidates for supporting an imminent handoff of a MN's traffic. This determination should be based upon the admission control status provided by the CTA associated with each AR in the context group.

Since the HODP is a functional element that deals primarily with the Handoff of micro-flows, it is not a part of the context transfer framework, per se. However, it is a primary collaborator in context transfer, and it is required to describe how the various context transfer scenarios would transpire.

5.2 CTA-HODP interface

A few transactions between the CTA and the HODP functions are necessary to discover a suitable handover candidate for the MN's traffic. This requires an interface definition between the two entities. The entity HODP can be co-located with the CTA in which case the CTA-HODP interface could have different implementations.

However, if the HODP is an entity that is not located in the same physical entity as the CTA then the CTA-HODP interface should be a standard handover candidate discovery protocol.

Since HODP is an entity outside context transfer (lies within the mobility framework), the details of this interface are outside the scope of this paper.

6. Additional features supported by the framework

6.1 Progressive context transfer

The proactive context transfer allows the transfer of the context for the MN to the context group members before an actual handover is required. During the time period from when the context is transferred proactively and the actual handover of the MN's microflows to any of the context group members occur; the context at the context transfer source (CTS) may get changed. These changes may include the addition or deletion of one or more microflows from the MN. The context information for the MN is composed of the configuration and state components (see Section 3). The intent of the proactive transfer is to prepare the potential context group candidates by replicating the information required to perform an admission control at each of the candidates. The state component, information that is updated on each packet arrival at the source, plays no role in the admission decision.

By separating the context into configuration and state components, context transfer can be performed in a progressive way. For a proactive context transfer mode, the information required to perform the admission decision (i.e. the configuration components) should be transferred proactively. Moreover, since this information changes infrequently, it should be kept synchronized between the CTS and the rest of the context group members. The events to trigger an update from the CTS to the rest of the context group members are simply the addition and deletion of microflows to the overall context associated with the MN's traffic at the CTS. The second stage of the context transfer, updating the state components, should be coupled with the actual handover. The context transfer here carries the information that is updated on each packet arrival and is now required at the new AR supporting the MN's traffic. This is effectively a one-to-one transfer, as only the new serving AR requires this information.

The nature of the information to be transferred as the context makes progressive context transfers a very useful concept. For example, the information contained in the state context is time-critical and needs to be transferred instantly within a single transfer but it does not play any role in the admission control. On the other hand, the configuration context is needed for performing an admission decision on the capabilities of the candidate AR but has no real-time significance. Hence coupling state context transfer with handover and performing configuration context transfer proactively meets the requirements of each of these information categories.

The usefulness of the state information transfer is dependent on the type of the feature itself. For example, the metering information on QoS type of context may not be very useful if it is not received before the first packet is forwarded at the candidate AR. A comprehensive study of each type of context is required to identify the information that could be useful as part of context. There are some documents that attempt to analyze various type of context information; [CT-QoS] [CT-HC] [QoS-REL] [CT-IPSec] attempt to capture the specific requirements posed by some of the context types.

6.2 Demand and reservation states in the proactive context transfer

The proactive transfer of configuration context and the outcome of the admission control on this context determine the level of support for the MN's traffic if the MN's traffic is re-directed to the AR. Since the transfer of the context to an AR and actual handover of the traffic to the AR may not happen at the same time, the resource availability at the AR may not be the same as it was reported after the admission control process on transferred context.

The framework provides concepts by which the mobility mechanism can make effective use of the admission control information and the reported status of the resource availability. This is accomplished by allowing the mobility mechanism to put the AR into one of the two states in terms of resources: the demand and the reservation states.

In the demand state, the AR has essentially acknowledged that it has the capability to support the MN's traffic, but no attempt is made to reserve resources. Thus, when the handover is finally attempted, the necessary resources may not be available and the handover to the AR may fail partially or completely. This option minimizes resource use, and requires some level of over-provisioning to ensure an acceptably low level of failed handovers. This could be used for the services that may not require immediate resource availability. Examples could be http sessions etc.

In the reservation state, an AR not only acknowledges that it has the capability to support the MN's traffic, it also reserves the necessary resources. In this case, when the handover is finally attempted, it should succeed. This option minimizes handover failures, at the expense of having resources allocated for potential handovers that may never occur. This could be useful for services that cannot tolerate disruptions like VoIP etc. The reservation mode guarantees the availability of resources for such services immediately after the handover.

7. Example scenarios based on the context transfer framework

This section provides some scenarios to explain the inter-working of the various functional entities and the corresponding protocols of the framework. Some mnemonics are used for the sake of brevity:

- MN54 is a moving mobile node in the access network. Initially, MN54 does not have any active microflows through any of the access routers but is within the coverage area of AR1.
- AR1 is an access router in the network and CTA1 is the context transfer agent at the AR1.
- AR7 is another access router in the same access network and CTA7 is the context transfer agent at AR7. At some point in time, AR7 receives a MADE notification of the presence of the MN54 in its coverage area due to the MN's mobility.
- MCDF is the membership collection and distribution function.
- HODP is an entity within the mobility framework that interacts with the CTA within context transfer framework.

7.1 Context group management scenarios

The addition and deletion of ARs to the context group, tracking the source(s) of the context group associated with each active MN in the network requires a context group management mechanism in place.

This mechanism also distributes the identity of the context source to the CGCs, which helps the CGCs initiating a context transfer to send a request directly to the source of the context. The following scenarios explain CG management in the context transfer framework (see Figure 3).

7.1.1 Joining the CG

1. AR1 receives a mobile arrival notification (MADE) for MN54. AR1 now becomes the context group candidate for MN54's context.
2. CTA1 informs the MCDF that AR1 is the CGC for MN54's context group, using MDP.
3. Assuming that MCDF does not have any existing CG information for MN54, it informs CTA1 that no context transfer source is available, using MDP.

7.1.2 Context and Context Group creation

4. MN54 establishes one or more microflows through AR1. CTA1 collects and stores the context associated with MN54's active microflows.
5. CTA1 informs the MCDF that it is supporting the active microflows for MN54 (and thus becomes the source of the MN54's context), using MDP.
6. MCDF creates a context group 'CG54' for the MN54's context withAR1 as the context transfer source (CTS).

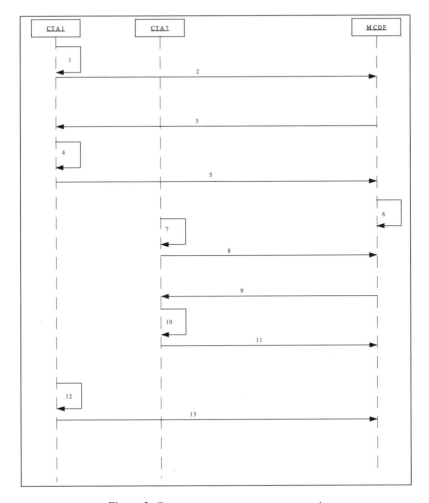

Figure 3. *Context group management scenarios*

7.1.3 Membership distribution

7. AR7 receives a mobile arrival notification (MADE) for MN54. AR7 now becomes a context group candidate for MN54's context.

8. CTA7 informs the MCDF that AR7 is a CGC for MN54's context group (OR MCDF receives a request from CTA7 for joining the CG associated with MN54), using MDP.

9. The MCDF informs CTA7 of the following information, using MDP:

- list of sources (CTS IDs) for the MN's microflows (only CTA1 in this case);
- context group information such as context group identification.

7.1.4 Updating the MCDF as the source for a CG changes

10. AR7 receives a change of forwarding path for MN54's traffic from HODP.

11. Being the new source for MN54's context, CTA7 informs the MCDF that it is supporting the active microflows for MN54, using MDP; if CTA1 still supports one or more microflows associated with the MN54's traffic, it will continue as one of the sources for the context within the CG associated with MN54.

7.1.5 Leaving the CG

12. AR1 determines that the MN54 is leaving its communication area through a mobile departure event (MADE):

- since AR1 is not the source of the context within the context group but just a context group member (replicating the context), it will only have to inform the source of the context to discontinue any updates. No update to the MCDF is required in this case.

13. If AR1 receives the departure event while it is hosting one or more microflows for MN54 (acting as a CTS), AR1 needs to inform MCDF that it is no more a source of the context group associated with MN54, using MDP.

7.2 Context transfer scenarios

The context transfer is initiated from the CGC to the CTS when the newly joined context group member has already retrieved the CTS information through the context group management mechanism.

- CTA1 is supporting the MN54's microflows and hence is the CTS within the CG for MN54;
- CTA7 is a CGC and it receives the CTS information from the MCDF.

7.2.1 Proactive context transfer mode (Figure 4)

7.2.1.1 Configuration context transfer

1. CTA7 receives the CTS information for MN54 from the MCDF.

2. CTA7 sends a context transfer request to CTA1 for MN54's configuration context, using CTP.

3. CTA1 transfers the configuration context for MN54, using CTP.

7.2.1.2 Context processing and inter-working with mobility mechanism

4. CTA7 processes the configuration context and determines the available support for MN54's traffic.

5. CTA7 informs the HODP about the available support for the MN54's traffic, using the CTA-HODP interface.

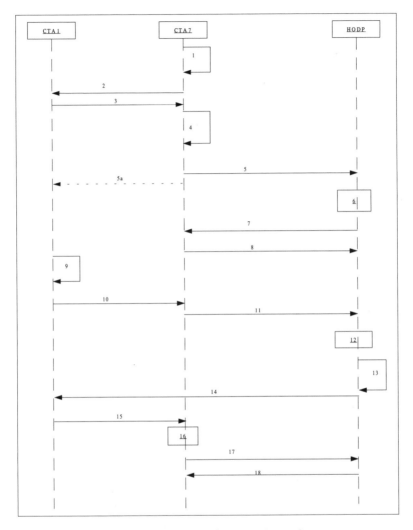

Figure 4. *Context transfer: proactive mode*

5a. If CTA7 is unable to admit MN54's traffic, it may opt to remain as a CGC but discontinue receiving the context updates from the source, using CTP.

6. The HODP determines demand and reservation states for the MN54's microflows. It may choose all the microflows for either demand or reservation states OR it may select some microflows under demand state and others in reservation state. For the microflows in demand mode, the HODP only records the current available support at the AR7. AR7, therefore, makes no attempt to reserve the resources for MN54's traffic.

7. In reservation mode, HODP requests CTA7 to reserve the resources to support MN's microflows, using the CTA-HODP interface.

8. CTA7 reserves the resources and sends the confirmation to the HODP, using the CTA-HODP interface.

7.2.1.3 Configuration context update

9. MN54 initiates a new microflow through CTA1 at some time while handover of the MN54's traffic has not yet performed to any of the context group members. The new microflow is added to the MN's context at the CTA1.

10. CTA1 sends an update on the configuration context of the new microflow to the context group members (only CTA7 in this case), using CTP.

11. The available support at AR7 for the new microflow is conveyed to the HODP, using the CTA-HODP interface.

12. The HODP again determines a mode (demand or reservation) for the newly added microflow.

7.2.1.4 State context transfer

13. The AR7 determines that the MN54 is moving away from its coverage area and initiates a handover event (HOE) to the HODP.

14. The HODP decides to handover the MN54 traffic to the CTA7 (best available CGC) and send a handover decision message to the CTA1.

15. CTA1 transfers the state context associated with the MN54's traffic to the CTA7, using CTP.

16. For demand state microflows, the CTA7 processes the current configuration context and determines the available support for MN54's microflows in demand state.

17. CTA7 informs the HODP about the available support for the MN54's traffic in demand state, using the CTA-HODP interface.

18. For a complete support available at the CTA7, the HODP acknowledges the information, using the CTA-HODP interface:

   - for a scenario where partial or no support was available at the CTA7 or CGC, the HODP must take necessary actions to redirect the MN54's traffic to a suitable handover candidate. Since HODP belongs to the mobility mechanism, the discussion of partial or complete failure of the admission control is out of the scope of this discussion.

7.2.2 Reactive context transfer mode (Figure 5)

1. The HODP receives a handover event (HOE) that the MN is being handover to AR7.

2. The HODP sends a message to the CTA1 to transfer the context associated with the MN54's traffic to CTA7, using the CTA-HODP interface.

3. CTA1 transfers the complete context (both configuration and state context) associated with the MN54's traffic to the CTA7, using CTP.

4. CTA7 processes the configuration context and determines the available support for MN54's traffic.

5. CTA7 informs the HODP about the available support for the MN54's traffic, using the CTA-HODP interface.

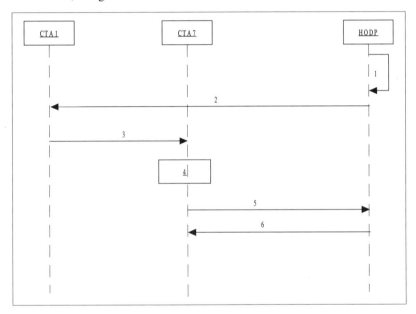

Figure 5. *Context transfer: reactive mode*

6. For a complete support available at the CTA7, the HODP acknowledges the information, using the CTA-HODP interface:

-   for a scenario where partial or no support was available at the CTA7; or
-   CGC, the HODP must take necessary actions to redirect the MN54's traffic to a suitable handover candidate. Since HODP belongs to the mobility mechanism, the discussion of partial or complete failure of the admission control is out of the scope of this contribution.

8. Conclusion

In a mobile environment, the context transfer between the nodes is required to preserve the service level provided to the end user or application. This paper proposes a framework for performing the context transfer between the Access Routers in a wireless access network. The framework identifies the functional roles of various elements that enable the context transfer. The required protocols between the functional entities are discussed with clear scenarios. The framework also discussed a close interaction of the context transfer framework with the mobility mechanisms.

9. Future work

The next step in this work is the definition of the context transfer protocol (CTP) and the Membership Distribution Protocol (MDP). It would also be useful to define the characteristics of the Mobile Arrival-Departure Event (MADE), handover event (HOE), the requirements for the HandOver Decision Point (HODP) and the CTA-HODP interface (required for the handover candidate discovery).

REFERENCES

[DES] Design Team, "Context Transfer: Problem Statement", June 2001. http://www.ietf.org/internet-drafts/draft-ietf-seamoby-context-transfer-problem-stat-01.txt

[SYE] SYED, H., KENWARD, G., et al, "General Requirements for a Context Transfer Framework", June 2001 (work in progress). http://www.ietf.org/internet-drafts/draft-ietf-seamoby-ct-reqs-00.txt

[CNP] NEUMILLER, P., "Contract Net Protocol". Internet draft.

[CT-802.11] ABOBA, B., et al, "A Model for Context Transfer in IEEE 802", July 2001. http://www.ietf.org/internet-drafts/draft-aboba-802-context-00.txt

[CT-ICMP] KOODLI, R., et al, "A Context Transfer Framework for Seamless Mobility", February 2001 (work in progress). draft-koodli-seamoby-ctv6–00.txt

[CT-HC] KENWARD, G., " Context Transfer Considerations for ROHC", July 2001 (work in progress). http://www.ietf.org/internet-drafts/draft-kenward-seamoby-ct-rohc-reqs-00.txt

[CT-QoS] SYED, H., MUHAMMAD, J., "QoS (DiffServ) Context Transfer", June 2001 (work in progress). http://search.ietf.org/internet-drafts/draft-hamid-seamoby-ct-qos-context-00.txt

[QoS-REL] WESTPHAL, C., PERKINS, C., "Context Relocation of QoS Parameters in IP Networks", July 2001 (work in progress). http://www.ietf.org/internet-drafts/draft-westphal-seamoby-qos-relocate-00.txt

[CT-IPSec] HAMER, L.-N., et al, "IP Sec Context Transfer", June 2001 (work in progress). http://search.ietf.org/internet-drafts/draft-hk-seamoby-ct-ipsec-00.txt

Chapter 2

Dynamic proxy selection for mobile hosts

Tsan-Pin Wang

Department of Computer Science and Information Management, Providence University, Taiwan, ROC

Lu-Fang Wang and Chien-Chao Tseng

Department of Computer Science and Information Engineering, National Chiao-Tung University, Taiwan, ROC

1. Introduction

The World Wide Web (WWW) is the most popular information dissemination system on the Internet. Accessing a Web server may be costly over the wide-area Internet because of the huge volume of traffic. Therefore, a client host may access the Web through a Web proxy in order to reduce the waiting time. A *Web proxy* is a Web caching server that caches popular Web documents to reduce Web accessing time [BAE 97, WES 98, GLA 94, LUO 94, BOL 96]. The first widely available Web proxy was the CERN Web proxy server developed in early 1994 [LUO 94]. According to [WES 98, WES 99], the Squid Web proxy server has become the most widely used Web proxy in recent years.

The advance of wireless technologies has made it feasible to access the Internet over wireless media. The Wireless Web is the World Wide Web (WWW) over the wireless Internet or other wireless networks. The wireless Internet poses new issues for Web browsing and Web-based applications. One issue is that wireless communication links are typically slow, expensive, and unreliable. Several studies have proposed various solutions. From most of the research [KAS 00, FOX 98, FOX 96, FLO 98, HOU 96, CHA 97] have developed Web proxies that filter out unnecessary data for mobile hosts before these data pass through the expensive wireless link. Santos [SAN 98] has proposed reduced vector graphics and animation contents that can replace the costly image pictures to pass through low-speed wireless links. Jiang and Kleinrock [JIA 98] designed a Web prefetching scheme that deals with the slow access speed of the wireless Web. Mazer and Brooks [MAZ 98] developed a disconnected update approach that handles the unreliable wireless link. Their approach supports the updating or creation of Web materials while the mobile host is disconnected, and supports the propagation back to the appropriate Web servers upon the reconnection of the mobile host.

In addition to the academic studies, an industry standard of the wireless Web has been proposed as well. The Wireless Application Protocol (WAP) [WAP 99a, WAP 99b, W3C 99] provides not only wireless Web accessing, but also Internet communications and advanced telephony services on digital mobile phones, pagers, personal digital assistants and other wireless terminals. The WAP caching model is studied in [WAP 99c].

The ideal position of Web proxies makes it possible to retain frequently requested documents and to keep them ready for subsequent requests by prospective different Web clients. Several effective proxy selection schemes have been proposed in the literature [WES 98, ROS 97, VAL 98, WES 97]. However, none of these schemes deal with the wireless environment. On the wireless Internet, a host may move around and retain the point of attachment to the Internet even while it is moving. Host mobility may affect the effectiveness of the Web proxy selection schemes proposed for the wired Internet. Consider an extreme case. A mobile host decides to access the Web through a proxy in Taiwan. Later, the mobile host moves to the USA for some purpose. While the mobile host is moving or even when the mobile host reaches the USA, it still accesses the Web through the fixed proxy in Taiwan. As a result, the waiting time may increase rather than decrease.

The goal of this paper is to help a mobile host to dynamically decide whether to use a fixed Web proxy or not. We propose a dynamic selection scheme. With this scheme, a mobile host may avoid suffering the unnecessary waiting time. We also propose a performance model to evaluate the proposed scheme.

The rest of this paper is organized as follows. Section 2 presents the background and related work. Section 3 describes the proposed dynamic selection scheme. Section 4 introduces the proposed performance model. Section 5 discusses the simulation results, comparing the selection schemes. Section 6 summarizes the conclusions.

2. Existing proxy selection schemes

This section presents the existing proxy selection schemes. There exist a few proxy selection schemes on the wired Internet. The existing schemes select a proper Web proxy among multiple candidate proxies for each connection. The most popular schemes among these are introduced in [WES 98, ROS 97]. The most popular schemes include the Internet cache protocol scheme and the hash routing scheme.

The Internet cache protocol (ICP) scheme [WES 98] selects a proxy server based on the hit/miss ratio and the round-trip time. The ICP scheme is primarily used between a proxy and a proxy. An example of the ICP-based-selection scheme can be found in [WES 97]. In contrast, the hash routing (HR) scheme [ROS 97] selects a proxy server based on a hash function. The HR scheme is used not only between a proxy and a proxy but also between a proxy and a Web client. An example of the hash-routing-based-selection scheme can be found in [VAL 98].

These schemes work well on the Internet. However, existing proxy selection schemes encounter problems when the environment changes to the wireless Internet. Both the ICP-based schemes and the HR-based schemes need a group of pre-defined candidate proxies. Usually, all these candidate proxies are close to each other. As a result, the mobile host may encounter a long waiting time when using a remote proxy. Consider the following scenario: Professor Tseng teaches computer science at NCTU (Taiwan) and he has a notebook as a mobile host. Thus, the mobile host's home network is at NCTU, and the mobile host may select a proxy from a group of NCTU proxies. Some day, this mobile host might move to UCLA (USA) because Professor Tseng attends a conference there. Although a group of candidate proxies is examined, the waiting time to pass through any NCTU proxies still becomes significant and makes these schemes unfeasible.

When all candidate proxies are close to each other, we can imagine that there is a virtual big proxy close to the home network. We will use the FPS scheme described later to represent this situation. Somebody may try to define a group of candidate proxies that are far away from each other. Through these candidates, he attempts to use the existing selection schemes on the wireless Internet. Unfortunately, the ICP-based schemes will waste time to probe too distant candidate proxies to be selected. In contrast, the HR-based schemes may select a far away proxy while proxies nearby are available. Furthermore, the mobile host may encounter access control problems as follows.

At the present time, if the mobile host wants to change to use a new proxy in the foreign network, the new proxy may probably refuse to serve for security reasons. This is due to the common access control policy, serving only hosts in the sub-network nearby. Follow the scenario above: Professor Tseng changes to use an UCLA proxy either by manual setting or by an existing selection scheme. But for security reasons, in some cases the UCLA proxy serves only hosts with local home addresses. Therefore, Professor Tseng's mobile host cannot be served because its NCTU home address though the mobile host has an UCLA care-of-address at that time. Hence, the mobile host might only either directly access the Web or access the Web through a few fixed proxies.

In the following, we describe two basic proxy selection schemes. The first basic scheme we will discuss is the *fixed proxy scheme* (FPS). With the FPS scheme, the mobile host always accesses the Web servers through a fixed proxy server.

Let us illustrate the fixed-proxy scheme with scenarios as in Figure 1. Suppose that the mobile host is powered on at the location 1, and it makes a Web request to access the Web server 1. This Web request misses the cache at the proxy server. So the access path passes through the proxy server, reaches the Web server 1, and returns vice versa. Since location 1 is near enough to the proxy server, the process is just like the typical Web proxy used for ordinary hosts.

In Figure 1, the mobile host moves away from the location 1 along the moving path and toward the location 5. When it comes to the location 5 that is far away from the proxy server, the mobile host makes the second Web request. Now the accessed Web server changes to the Web server 3. The Web request passes through a remote proxy server, happens to miss the cache, and reaches the Web server 3 that is relatively near to the current position of the mobile host. The access path becomes inefficient in this situation. Even if the Web request hits the cache at the remote proxy server, the access path is still inefficient.

This is what we call the long waiting time problem through a fixed proxy. This effect becomes more significant if the waiting time between the mobile host and the Web server is relatively much shorter than the waiting time between the mobile host and the remote proxy server. For example, we can imagine what will happen if the proxy server is in the US, and the mobile host and the Web server 3 are in Taiwan.

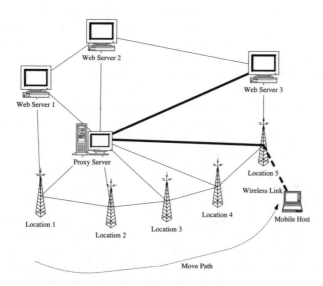

Figure 1. *The fixed proxy scheme (when the MH moves to location 5)*

In general, most mobile hosts exhibit *spatial locality*, i.e., mobile hosts tend to re-visit the locations that they visited previously. Consider the following scenario, the mobile host moves away from the location 5 along the moving path and turns back to location 2. Suppose that the mobile host moves by the location 4 and location 3 without making a Web request. When it comes to the location 2 that is near to the proxy server, the mobile host makes the third Web request. The accessed Web server this time is still the Web server 3. Fortunately, the Web request hits the cache at the proxy server nearby, and the waiting time of the mobile user is saved. The process is like the typical Web proxy used for ordinary hosts, again.

Another basic scheme we will discuss is the *direct scheme* (DS). With the direct scheme, the mobile host always directly accesses the Web servers without passing through a proxy server.

Consider the direct scheme as shown in Figure 2 with the same scenario of above examples. When the mobile host is at location 1, although the proxy server is near location 1, the Web server 1 is near location 1, too. Therefore, the waiting time of the mobile user is still acceptable.

When the mobile host moves to location 5, the waiting time between the mobile host and the Web server 3 is relatively much shorter than the waiting time between the mobile host and the proxy server. As a result, the waiting time experienced by the mobile user is more efficient in comparison to the FPS scheme. However, in Figure 2, the waiting time between the mobile host and the Web server 3 is much longer than the waiting time between the mobile host and the proxy server. Consequently, the mobile user may waste unnecessary waiting time if the Web request hits the cache with the FPS scheme.

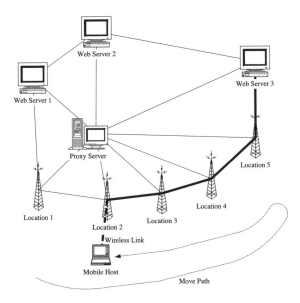

Figure 2. *The direct scheme (when the MH moves to location 2)*

3. Dynamic Selection Scheme

This section proposes our new scheme called the *dynamic selection scheme* (DSS).

3.1 Scheme description

The DSS works with the existing Web system. The design principle of DSS is described as follows. While the mobile host is moving, DSS performs the proxy selection scheme. If direct access is much better, DSS selects to directly access the Web server. If accessing the Web server through a proxy is much better, DSS selects to pass through the fixed Web proxy. If the two accessing methods are almost the same, DSS just keeps using the current accessing method.

In general, the DSS serves as a transparent proxy on the mobile host as shown in Figure 3. Conceptually, DSS operates like a cascading Web proxy to the specific Web proxy on the fixed host. To achieve the design principle, we should consider two important issues. The first issue is when to select a new method. When is the proper opportunity for the DSS to perform the selection? The second issue is how to select a better method. When the proper opportunity occurs, how does the DSS select a suitable accessing method? We will discuss these two issues in the next subsections.

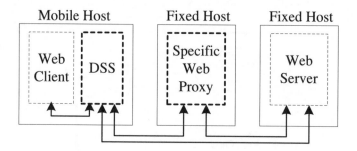

Figure 3. *Operation of the Dynamic Selection Scheme*

3.2 When to select: the decision phase

When is the proper opportunity for the DSS to perform the selection while the mobile host is moving? We derive the following formula to perform the selection (Figure 4). The first term is the mobility-to-request ratio. The numerator of this mobility-to-request term is the number of sub-nets that the mobile host strode across. This number of sub-nets is counted from the moment that the mobile host is powered on. The denominator of this mobility-to-request term is the number of requests that the mobile host launched. This number of requests is counted from the moment that the mobile host is powered on. The second term is the *ReqCount*.

ReqCount is the number of requests that is counted from the moment of the last selection till now. The decision will be made if the term on the left-hand side is more than a threshold value, *whenHP.*

$$\frac{\text{\# of sub-nets strode across}}{\text{\# of requests}} * ReqCount > whenHP$$

Figure 4. *Decision formula*

Using this formula, we develop the decision procedure of DSS. The DSS decides the opportunity with the procedure shown in Figure 5. This figure also shows the procedure of the real Web accessing action.

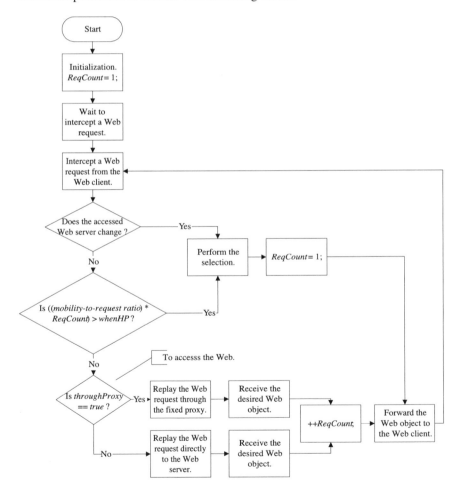

Figure 5. *Flowchart of the decision procedure*

3.3 How to select: the selection phase

When the proper opportunity occurs, how does the DSS select a suitable accessing method? The procedure by which the DSS selects a suitable accessing method is shown in Figure 6. In this procedure, the DSS program on the mobile host needs two parameters, the round-trip time (RT_{PS}) and the hit ratio (P_{Hit}) to assist the selection procedure. To offer the round-trip time and the hit ratio, the modified Web proxy that we call the specific Web proxy executes the procedure shown in Figure 7.

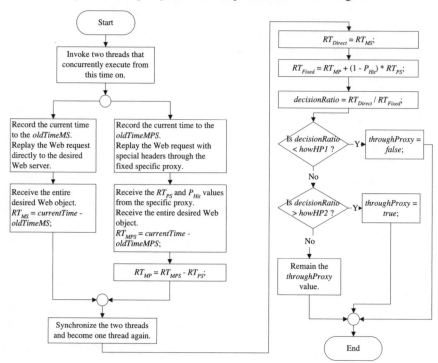

Figure 6. *Flowchart of the selection procedure*

4. Performance analysis

This section proposes a performance model to compare DS, FPS, and DSS.

4.1 Web round-trip time analysis

4.1.1 Performance metric

We follow the performance metrics proposed in [ROS 97] that has made a comparison among several proxy selection schemes on the wired Internet. For the wireless Internet, we consider the performance metric: average round-trip time to

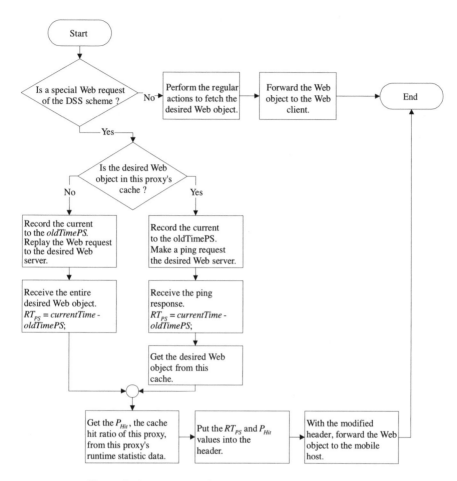

Figure 7. *Operation to collect parameters at a specific proxy*

satisfy a Web request for a Web object. The clock begins when the mobile host requests the object and ends when the mobile host receives the entire object. This round-trip time includes TCP connection establishment delays, propagation delays, transmission delays, server processing delays, queuing delays, TCP slow-start delays, and other delays for the wired part. The round-trip time also includes handoff delays, error recovery delays, and other delays for the wireless part. For simplicity, we ignore the DNS-lookup delay and local Web caching on the mobile host.

4.1.2 Web round-trip time model

We modified the analytic model developed in [ROS 97] to develop the Web round-trip time model. We assume that the environment is supported with the mobile IP software by the routing optimization mechanism.

Let RT_{MS} be the round-trip accessing time to request and transfer an object between the mobile host and the accessed Web server. The clock for this round-trip time begins when the mobile host requests the object and ends when the mobile host receives the entire object. The round-trip time includes all delays for the wired part and all delays for the wireless part mentioned in the previous section. Let RT_{MP} be the round-trip accessing time between the mobile host and the single fixed proxy. Let RT_{PS} be the round-trip accessing time between the single fixed proxy and the accessed Web server.

The round-trip time for the direct scheme, T_{Direct}, is

$$T_{Direct} = RT_{MS} \qquad \text{Eq. 1}$$

Also, P_{Hit} is the hit probability of the proxy. The round-trip time for the fixed-proxy scheme, T_{Fixed}, is

$$T_{Fixed} = RT_{MP} + (1 - P_{Hit})\, RT_{PS} \qquad \text{Eq. 2}$$

No matter which scheme is used, the traffic passes through the same wireless link. Consequently, all schemes suffer similar handoff delays, error recovery delays, and other delays in the wireless environment. For simplicity, we ignore the handoff delays, the error recovery delays, and all other delays for the wireless part. In the following, we focus on modeling the mobility property of the mobile host.

The delay time to pass through a sub-net is defined as a constant time c. If we don't take the traffic congestion into account, the router processing delay dominates the network transmission time. The delay time within a sub-net has been modeled in Chapter 3 of [BER 92]. Thus, the Web round-trip delay time can be estimated as multiplying the number of sub-nets on a route path by the c.

Let d_{MS} be the distance (the number of sub-nets on the route path) between the mobile host and the accessed Web server. Let d_{MP} be the distance (the number of sub-nets on the route path) between the mobile host and the single fixed proxy. Let d_{PS} be the distance (the number of sub-nets on the route path) between the single fixed proxy and the accessed Web server. Eq. 1 becomes

$$T_{Direct} \approx d_{MS} \cdot c + d_{MS} \cdot c = 2 \cdot d_{MS} \cdot c \qquad \text{Eq. 3}$$

Also, Eq. 2 becomes

$$T_{Fixed} \approx 2 \cdot d_{MP} \cdot c + (1 - P_{Hit}) \cdot (2 \cdot d_{PS} \cdot c) \qquad \text{Eq. 4}$$

Taking the expected values on Eq. 3 and Eq. 4, we can derive that the average T_{Direct} is

$$E[T_{Direct}] = E[2 \cdot d_{MS} \cdot c] = 2 \cdot c \cdot E[d_{MS}]$$ Eq. 5

Also, the average T_{Fixed} is

$$E[T_{Fixed}] = E[2 \cdot d_{MP} \cdot c + (1 - P_{Hit}) \cdot (2 \cdot d_{PS} \cdot c)]$$ Eq. 6

$$= (1 - P_{Hit}) \cdot (2 \cdot d_{PS} \cdot c) + 2 \cdot c \cdot E[d_{MP}]$$

4.2 Average distance analysis

In this section, we propose the model to derive $E[d_{MS}]$ and $E[d_{MP}]$. Four dominant factors will heavily affect the average distances and the average round-trip times between host pairs as follows:

1. What is the arrival pattern of the Web requests?
2. What is the movement pattern of the mobile host?
3. What is the spatial locality pattern of the mobile host?
4. Where are the accessed Web servers?

We will derive these factors using following models.

4.2.1 Web request model

Pitkow [PIT 98] summarized that the arrival process is bursty, similar between the micro second and minute time range, and the arrival process has periodic patterns which are able to be modelled by time series analysis in the hour to weekly time range. Judge et al. [JUD 97] postulated a model for HTTP request traffic with lognormally distributed inter-arrival times and an invariant shape parameter of 1.0. They also stated that a simulation study indicated that the estimate of response time performance for a WWW proxy server is insensitive to the choice of their model of log-normally distributed HTTP request inter-arrival times in place of the Poisson arrival process assumption. Based on this result, we chose the Poisson arrival process as our HTTP request arrival process.

4.2.2 The MH mobility model

4.2.2.1 Applying the two-dimension random walk model

Lin and Mak [LIN 94] proposed a two-dimension random walk simulation model to study the mobility pattern of the mobile host. The random walk model is simple and suitable to use. Thus, we also use a two-dimension random walk simulation model.

As shown in Figure 8, a square represents a sub-net and a sub-net has four neighbors. The up, down, left, and right squares are neighbors to which the mobile host has equal probability to move. The route paths are determined by using this random walk model. For comparison, we assume that mobile IP has routing optimization mechanism. Therefore, the route path between two sub-nets (two squares) is the straight line between these two squares.

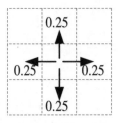

Figure 8. *Two-dimension random walk*

4.2.2.2 Applying the request-to-mobility model

Lin [LIN 97] derived the probability $\alpha(K)$ that a portable moves across K registration areas (RAs) between two phone calls. The arrival process of incoming phone calls is assumed to be the Poisson process in [LIN 97]. Our model adopts the assumptions of an α function in [LIN 97]. Let $\alpha(D)$ be the probability that a mobile host moves to new sub-nets D times between two consecutive outgoing requests. The concept of $\alpha(D)$ is used to integrate the Web-request model and the MH mobility model. The integrated model is shown in Figure 10.

Let λ_r be the request arrival/outgoing rate and λ_m the mobile host mobility rate. We define V as the variance of the mobile host residence time and $\theta = \frac{\lambda_r}{\lambda_m}$ the request-to-mobility ratio. We assume that the outgoing HTTP requests are a Poisson process and the time the mobile host resides in a sub-net has a general distribution. Further, we assume that the mobile host residence time has a Gamma distribution. We obtain that $f_m^*(s)$, the Laplace-Stieltjes transform of the mobile host residence time in a sub-net, is equal to

$$f_m^*(s) = \left(\frac{\lambda_m \cdot \gamma}{s + \lambda_m \cdot \gamma}\right)^\gamma, \gamma = \frac{1}{V \cdot \lambda_m^2}.$$

Eq. 7

From [LIN 97] and [LIN 98], we obtain that

$$\alpha(D) = \begin{cases} 1 - \dfrac{1 - f_m^*(\lambda_r)}{\theta}, D = 0 \\[2mm] \dfrac{1}{\theta} \cdot [1 - f_m^*(\lambda_r)]^2 \cdot [f_m^*(\lambda_r)]^{D-1}, D > 0 \end{cases},$$

Eq. 8

4.2.3 MH locality model

Lin and Tsai [LIN 98] usedà(K) and a two-dimensional random walk with reflecting barriers to model the portable's locality behavior. We follow this idea and use a two-dimension random walk with reflecting barriers to model the spatial locality of a mobile host as shown in Figure 9.

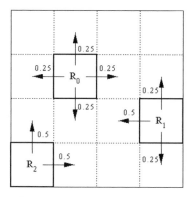

Figure 9. *Two-dimensional random walk with reflecting barriers*

Without loss of generality, we assume that the home network is at the center of the locality area of the mobile host, and the proxy (if exists) is at the center of the locality area of the mobile host. The result of the derived simulation environment is shown in Figure 10.

4.2.4 Web server distribution model

For simplicity, there is one accessed Web server shown in our simulation model as in Figure 10. We can control the position of the accessed Web server, and observe the effect of the accessed-Web-server position to the performance. It is obvious that the case with multiple Web servers can work well in the current simulation model.

4.3 Simulation procedures

In this section, we evaluate the average round-trip time for each scheme by simulation. The simulation steps for the DS scheme are described in Figure 11. On the other hand, the simulation steps for the FPS scheme are described in Figure 12. In a similar manner, the simulation steps for the DSS scheme are similar to Figures 11 and 12. Combining the steps of the DSS shown in Figures 4, 5, 6 and 7, we can easily simulate the average round-trip time of the DSS scheme.

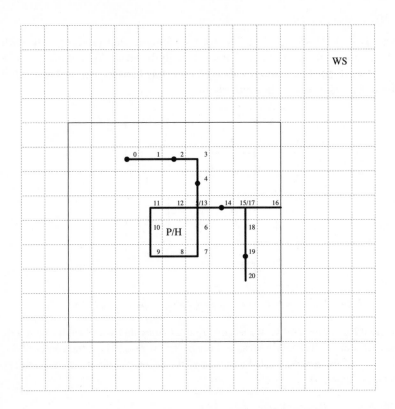

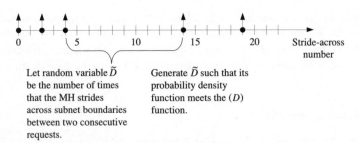

Let random variable $\tilde{D}$ be the number of times that the MH strides across subnet boundaries between two consecutive requests.

Generate $\tilde{D}$ such that its probability density function meets the (D) function.

Figure 10. The simulation environment

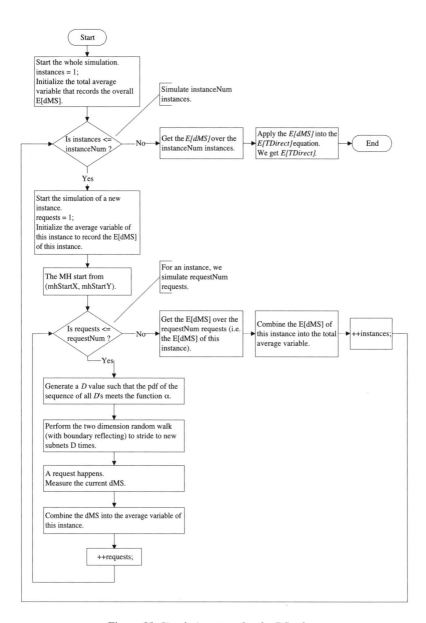

Figure 11. *Simulation steps for the DS scheme*

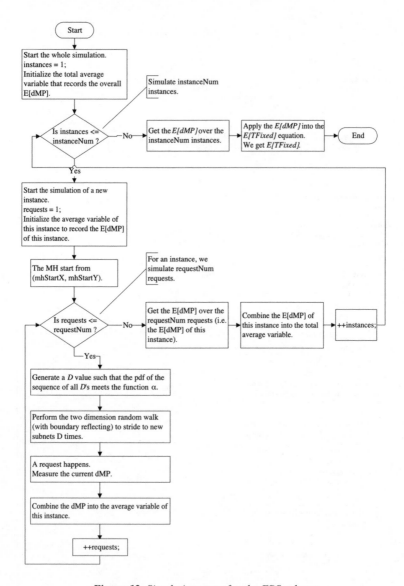

Figure 12. *Simulation steps for the FPS scheme*

5. Simulation results and discussion

This section compares DS, FPS, and DSS based on our model and simulations.

The effect of topology is shown in Figure 13. In this figure, the DSS scheme is always much better than the worst scheme and is always just a little worse than the best scheme in all circumstances.

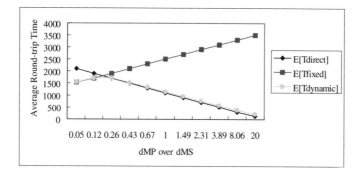

Figure 13. The effect of topology

The effect of the hit ratio is shown in Figure 14. The horizontal axis is the hit ratio, and the vertical axis is the average round-trip time. This figure shows that the DSS scheme is always much better than the worst scheme and is always just a little worse than the best scheme in all circumstances.

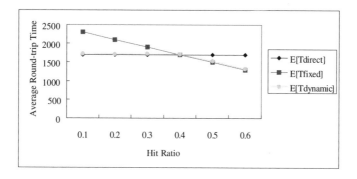

Figure 14. The effect of the hit ratio

According to Figures 13 and 14, the DSS scheme is always much better than the worst scheme and is always just a little worse than the best scheme in all circumstances. With the DSS scheme, the mobile user need not have the luck to always choose the best scheme while the mobile host is moving. Moreover, the mobile user need not manually change the Web proxy setting frequently. The DSS scheme can effectively reduce the potential long waiting time for the mobile user.

The effect of the mobility-to-request ratio is shown in Figure 15. The horizontal axis is the mobility-to-request ratio, and the vertical axis is the average round-trip time.

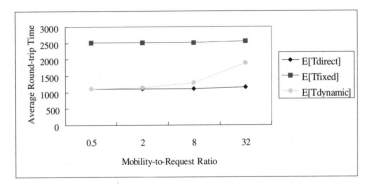

Figure 15. *The effect of the mobility-to-request ratio*

Figure 15 reveals a potential problem for the DSS scheme. The overhead on the delay time increases as the mobility-to-request ratio grows. Let us observe the impact of *whenHP* values on this problem. Figure 16 shows that the potential problem can be controlled by *whenHP*. The mobile user can restrain the increase of overhead on the delay time by setting a *whenHP* value according to the user's usual using and moving habits. This benefits the mobile user since the number of times needed to change the *whenHP* setting is much less than the number of times to change the Web proxy setting.

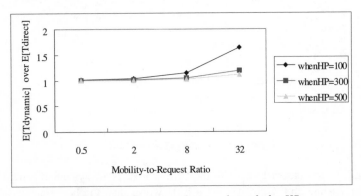

Figure 16. *Recommendation values of whenHP*

The potential long waiting time may increase if the DSS is too insensitive to perform selections. However, this potential problem can be controlled by selection of a suitable *howHP* pair as shown in Figure 17. The mobile user can restrict the increase in the potential long waiting time by setting a *howHP*-value pair according to the user's habits and need.

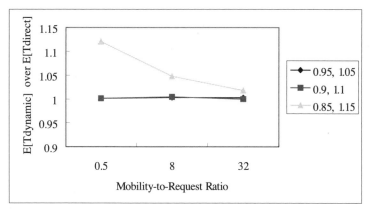

Figure 17. Recommendation values of howHP

6. Conclusions

This paper proposes a new scheme of Web-proxy selection that reduces the potential long waiting time of the Web access for mobile computing. The proposed proxy selection scheme can be applied to dynamically decide whether to use a fixed Web proxy or not when a mobile host visits a new network. The DSS scheme consists of two phases: a decision phase and a selection phase. The decision phase uses a heuristic function to determine when to trigger the selection phase to choose a better accessing path. The penalty has been shown to be the overhead on the delay time and on the network traffic in order to measure the round-trip times. The DSS scheme is designed to be concise and compact. Therefore, the DSS scheme is suitable to be implemented on lower computing-power mobile hosts.

We conducted a performance evaluation on the proposed dynamic selection scheme. The DSS scheme is compared with the two basic access methods, direct accessing and Web accessing through fixed proxies, in terms of the average access time. According to the simulation results, the DSS scheme is always much better than the worst scheme and is always just a little worse than the best scheme in all circumstances. With the DSS, the mobile user need no t have the luck to always choose the best scheme while the mobile host is moving. With the DSS, the mobile user need not manually change the Web proxy setting frequently either. The DSS scheme can effectively reduce the potential long waiting time for the mobile user.

A potential problem for the DSS is that the overhead on the delay time increases as the mobility-to-request ratio grows. Fortunately, this effect can be controlled by a threshold value, *whenHP*, which is an adjustable parameter. The mobile user can restrain the increasing of the overhead on the delay time by setting a *whenHP* value according to the user's usual using and moving habits. This benefits the mobile user, since the number of times needed to change the *whenHP* setting is much less than the number of times to change the Web proxy setting if the user does not use the DSS scheme at all. Furthermore, we have provided recommended values of the *whenHP* values for the mobile user to refer to.

Another potential problem for the DSS is that the potential long waiting time increases if the DSS is too insensitive to perform selections. However, the *howHP* pair, an adjustable-parameter pair of the DSS, can control this effect. The mobile user can restrict the increase of the potential long waiting time by setting a *howHP*-value pair according to the user's habits and need. Moreover, we have offered recommended values of the *howHP*-value pairs for the mobile user to refer to.

There is another application for the DSS scheme. We may combine the concepts within the DSS, the concepts of mobile computing, into the existing Web-proxy-selection schemes for the wired Internet. As a result, we may improve the existing Web-proxy-selection schemes such that these schemes turn into being appropriate for the wireless Internet.

Acknowledgement

This research was supported by the National Science Council, ROC, under grant NSC 89-2213-E-126-010.

REFERENCES

[BAE 97] M. Baentsch, L. Baun, G. Molter, S. Rothkugel, and P. Sturn. "World Wide Web Caching: The Application-Level View of the Internet", *IEEE Communications Magazine,* Vol. 35, No. 6, pp. 170–178, June 1997.

[WES 98] D. Wessels and K. Claffy. "ICP and the Squid Web Cache", *IEEE Journal on Selected Areas in Communications,* Vol. 16, No. 3, pp. 345–357, April 1998.

[GLA 94] S. Glassman. "A Caching Relay for the World Wide Web", *Proceedings of the First International World Wide Web Conference,* May 1994.

[LUO 94] A. Luotonen and K. Altis. "WWW Proxies", *Proceedings of the First International World Wide Web Conference,* May 1994.

[BOL 96] J.-C. Bolot and P. Hoschka. "Performance Engineering of the World Wide Web: Application to Dimensioning and Cache Design", *Proceedings of the Fifth International World Wide Web Conference,* May 1996.

[WES 99] D. Wessels. "Squid Internet Object Cache", Squid Project Home Page, June 1999, http://squid.nlanr.net/Squid/

[KAS 00] A. Kassler, A. Neubeck, and P. Schulthess, "Filtering Wavelet Based Video Streams for Wireless Interworking", *IEEE International Conference on Multimedia and Expo,* Volume 3, pp.1257–1260, 2000.

[FOX 98] A. Fox, I. Goldberg, S.D. Gribble, D.C. Lee, A. Polito, and E.A. Brewer. "Experience With Top Gun Wingman, A Proxy-Based Graphical Web Browser for the USR PalmPilot", *Proceedings of the IFIP International Conference on Distributed Systems Platforms and Open Distributed Processing* (Middleware '98), September 1998.

[FOX 96] A. Fox and E.A. Brewer. "Reducing WWW Latency and Bandwidth Requirements via Real-Time Distillation", *Proceedings of the Fifth International World Wide Web Conference*, pp. 48, May 1996.

[FLO 98] R. Floyd, B. Housel, and C. Tait. "Mobile Web Access Using eNetwork Web Express", *IEEE Personal Communications*, Vol. 5, No. 5, pp. 47–52, October 1998.

[HOU 96] B.C. Housel and D.B. Lindquist. "WebExpress: A System for Optimizing Web Browsing in a Wireless Environment", *Proceedings of the Second Annual International Conference on Mobile Computing and Networking (MOBICOM '96)*, pp. 108–116, 1996.

[CHA 97] H. Chang, C. Tait, N. Cohen, M. Shapiro, S. Mastrianni, R. Floyd, B. Housel, and D. Lindquist. "Web Browsing in a Wireless Environment: Disconnected and Asynchronous Operation in ARTour Web Express", *Proceedings of the Third Annual ACM/IEEE International Conference on Mobile Computing and Networking (MOBICOM '97)*, pp. 260–269, 1997.

[SAN 98] L. Santos. "Multimedia Data and Tools for Web Services over Wireless Platforms", *IEEE Personal Communications*, Vol. 5, No. 5, pp. 42–46, October 1998.

[JIA 98] Z. Jiang and L. Kleinrock. "Web Prefetching in a Mobile Environment", *IEEE Personal Communications*, Vol. 5, No. 5, pp. 25–34, October 1998.

[MAZ 98] M.S. Mazer and C.L. Brooks. "Writing the Web While Disconnected", *IEEE Personal Communications*, Vol. 5, No. 5, pp. 35–41, October 1998.

[WAP 99a] Wireless Application Protocol Forum Ltd. "WAP Forum Home Page", June 1999, http://www.wapforum.org/

[WAP 99b] Wireless Application Protocol Forum Ltd. "WAP Forum Specifications", June 1999, http://www.wapforum.org/what/technical.htm

[W3C 99] World Wide Web Consortium. "W3C: Mobile Access", June 1999, http://www.w3.org/Mobile/

[WAP 99c] Wireless Application Protocol Forum Ltd. "User Agent Caching", February 1999, http://www.wapforum.org/what/technical1_1/UACaching-11-Feb-1999.pdf.

[ROS 97] K.W. Ross. "Hash Routing for Collections of Shared Web Caches", *IEEE Network Magazine*, Vol. 11, No. 6, pp. 37–44, November–December 1997.

[VAL 98] V. Valloppillil and K.W. Ross. "Cache Array Routing Protocol v1.0", *IETF Internet Draft*, February 1998.

[WES 97] D. Wessels and K. Claffy. "Internet Cache Protocol (ICP), version 2", *IETF RFC 2186*, September 1997.

[BER 92] D. Bertsekas and R. Gallager. *Data Networks*, 2nd ed., Prentice Hall, 1992.

[PIT 98] J.E. Pitkow. "Summary of WWW Characterizations", *Computer Networks and ISDN Systems*, Vol. 30, pp. 551–558, April 1998.

[JUD 97] J. Judge, H.W.P. Beadle, and J. Chicharo. "Modeling World Wide Web Request Traffic", *Proc. SPIE – Int. Soc. Opt. Eng.*, Vol. 3020, pp. 92–103, February 1997.

[LIN 94] Y.-B. LIN and V.W. MAK. "Eliminating the Boundary Effect of a Large-Scale Personal Communication Service Network Simulation", *ACM Transactions on Modeling and Computer Simulation*, Vol. 4, No. 2, pp. 165–190, April 1994.

[LIN 97] Y.-B. LIN. "Reducing Location Update Cost in a PCS Network", *IEEE/ACM Transactions on Networking*, Vol. 5, No. 1, pp. 25–33, February 1997.

[LIN 98] Y.-B. LIN and W.-N. TSAI. "Location Tracking with Distributed HLR's and Pointer Forwarding", *IEEE Transactions on Vehicular Technology*, Vol. 47, No. 1, February 1998.

Chapter 3

An efficient simulation model for wireless LANs applied to the IEEE 802.11 standard

Paul Mühlethaler and Abdellah Najid

INRIA, Rocquencourt, Domaine de Voluceau, France

1. Introduction

Simulations for LANs were extensively presented in numerous papers in the 80s with the emergence of widely accepted standards such as Ethernet [4] or Token Ring [5]. The late 90s saw the emergence of a lot of standardization work for wireless LANs e.g. IEEE 802.11 [2], HiPERLAN [1], Bluetooth, etc. This opened up new business opportunities at the same time as wireless LANs raised new technical problems. At the medium access level, the main difference between LANs and wireless LANs lies in the fact that in LANs one usually has an atomic view of an event. This is not the same in wireless LANs due to the propagation effect. This leads to hidden nodes, capture effect or spatial reuse. See Figure 1.

Figure 1. Nodes A and B are simultaneously transmitting a packet. Node C is not able to decode either the packet sent by A or B while node D receives correctly the packet sent by node A

The first effect is usually considered negative for the network performance whereas the last two are usually presented in a positive way. However, all these phenomena are due to the same cause which is the high propagation decay of radio signals. In this paper, we give a simple but general model to take into account these effects and the behaviour of CSMA (Carrier Sense Multiple Access) techniques. Then, we show how the access scheme of the IEEE 802.11 can be modeled both precisely but also simply and efficiently. In fact it is intended to use this model to simulate ad-hoc networks and the MAC and physical layer must therefore be efficiently implemented in order to simulate complex scenarios over a reasonable duration. This paper is organized as follows: the next section describes the model of the physical layer. Section 3 first recalls the IEEE 802.11 access technique and then describes the model used to simulate this access protocol. Section 4 presents simulation results. This section begins by a detailed analysis of the IEEE 802.11 standard. Then the section gives performance evaluations of the IEEE 802.11 DS standard with a data rate of 1, 2, 5.5, 11 Mbit/s. We also study special behaviour such as broadcast transmission, performance with hidden nodes, spatial reuse.

2. Model for the physical layer

2.1 Linearity and propagation law

The main assumption of this paper is that we have a linear superposition of signals sent by potential transmitters. This model naturally leads to introducing a transmission matrix $cs_{i,j}$ which gives the strength of the signal sent by node j to node i. The signal strength $Pow(i)$ received by node i is therefore

$$Pow(i) = \sum_{j=1}^{n} a_j cs_{i,j}$$

where $a_j = 1$ if node j is transmitting or $a_j = 0$ otherwise.

Simple propagation laws of radio signals usually have the following expression

$$cs_{i,j} = \frac{P_j}{r_{i,j}^\alpha}$$

where
- P_j denotes the power sent by node j;
- r_{ij} denotes the distance between node i and node j;
- α denotes the signal decay, usually $2 \leq \alpha \leq 4$.

Of course this is an approximate model. However it should be noted that *the only important assumption* is the *linearity* of the model. We can actually use this linear model with all existing propagation models or pre-computed figures. All we will need is the transmission matrix $cs_{i,j}$.

2.2 Carrier sensing and reception

We now need to introduce the carrier sensing parameter. This parameter is a threshold above which the channel is assumed to be busy. In a CSMA protocol this threshold makes it possible to decide whether the channel is idle or busy. We will call this parameter the *carriersenselevel*.

In the previous section, we introduced the physical transmission model. Actually we need extra conditions to ensure the correct reception of packets.

We will assume that a packet sent by node i to node j in the transmission interval $[t_b, t_e]$ is correctly received by node j if

$$— \forall t \in [t_b, t_e] \quad cs_{i,j}(t) \geq datalevel$$

$$— \forall t \in [t_b, t_e] \quad \frac{cs_{i,j}(t)}{\sum_{k \neq j} a_k(t) cs_{i,k}(t)} \geq capturelevel$$

We have introduced two parameters: the *"datalevel"* and the *"capturelevel"*. The *"datalevel"* corresponds to the signal strength necessary to successfully transmit a signal. The *"capturelevel"* corresponds to the minimum value of a signal to noise ratio to successfully decode a transmission.

Actually, we will add an extra condition. We will assume that a correct reception can only start when a node receives a signal strength less than the carrier-sense level. This assumption implies that a correct transmission cannot start during a bad transmission. But conversely it is possible that a correct transmission starts and is corrupted by a new starting collision. If the second transmission starts significantly later than the first transmission then one usually calls this collision a late collision. In the case of correct operation of the carrier sensing, this can only occur with hidden node.

3. Model for the Medium Access Layer

3.1 The IEEE 802.11 MAC scheme

3.1.1 A CSMA technique

In this part, we will not address the centralized access mode called CF (Centralized Coordination Function) of the IEEE 802.11 standard. We will only deal with the distributed access scheme which, in the standard, is called the DCF (Distributed Coordination Function). This scheme is primarily based on a CSMA (Carrier Sense Multiple Access) scheme. The main principle of this access technique is a preventive listening of the channel to be sure that no other transmission is on the way before transmitting its packet. If the sensing of the channel indicates an ongoing transmission then the node waiting to start its transmission draws a random back-off delay. At the end of the outgoing transmission this back-off will be decremented whenever the channel is free (no carrier sensed). The node starts its transmission when its back-off delay reaches 0. This mechanism is presented in Figure 2. Nodes B and C receive a new packet to transmit while node A is transmitting. Node C draws the smallest back-off delay. The back-off delay of node C is decremented after the end of the transmission of node A. When the back-off delay of C expires, node C starts transmitting. After the end of C's transmission, B carries on decrementing its remaining back-off delay. When this back-off delay reaches 0, B starts transmitting. This scheme has often been called CSMA/CA where CA stands for Collision Avoidance. In fact, this terminology is quite approximate. It can notice however that this back-off strategy differs from the Ethernet strategy where the events on the channel are not taken into account to decrement the back-off delay; of course when the channel is still busy at the end of the back-off time the transmission attempt is made at the end of the carrier (1 persistence). Experts in medium access technique will identify a "tree" algorithm in the IEEE 802.11 CSMA/CA scheme. Indeed this protocol can be implemented with a collision counter and this counter is decremented on idle slots as is done for a "tree" or "stack" algorithm [6].

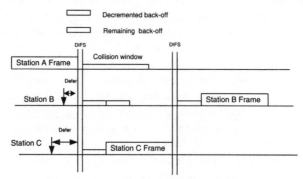

Figure 2. The IEEE 802.11 back-off mechanism

3.1.2 The MAC acknowledgement

With radio signals, it is not possible to directly detect collisions in a radio network. Indeed, it is not possible to listen to alien transmission while actually transmitting. Packet collisions must therefore be detected by another means. The IEEE 802.11 standard uses an acknowledgement for a point to point packet, broadcast packets are not acknowledged. This acknowledgement packet is sent by the receiver just after reception of the packet. The interframe between a packet and its acknowledgement (SIFS short interframe spacing) is shorter than between the end of a transmission and a packet (DIFS distributed interframe spacing). Therefore, the transmission of the acknowledgement will precede any other transmission attempt. Figure 3 shows how the acknowledgement works in the IEEE 802.11 standard.

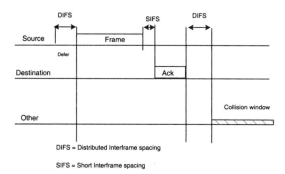

Figure 3. The IEEE 802.11 acknowledgement mechanism

3.1.3 The RTS/CTS and the NAV

In the previous subsection we have shown that it is impossible to directly detect collisions in wireless LANs. Therefore, when a collision occurs, the whole packet duration is lost. This reduces performance both in terms of achievable throughput and in delays. The RTS/CTS can cope with this problem. In this scheme, an RTS packet (Request to Send) is addressed by the source to the destination which, if the RTS has been well received, responds with a CTS packet (Clear To Send). If the source receives correctly the CTS packet, it will then transmit its packet to the destination. It is clear that the effect of a collision is reduced with this mechanism since for a collision only the RTS/CTS time is used. Moreover the RTS/CTS also has a very interesting effect on hidden nodes. This effect is obtained via the extra NAV (Network Allocation Vector) mechanism. Indeed the RTS and CTS packets hold the forthcoming transmission duration in the NAV field, see Figure 4. The RTS is therefore indicating to the source neighbors the duration of the transmission. This is of course interesting but in most cases this indication is indirectly available to the source's neighbors via the carrier sense. More interesting is certainly the effect of the CTS. All the destination's neighbors will then be aware of the forthcoming transmission duration. Some of them may be out of carrier sense reach from the source node; these nodes are potentially hidden nodes. Their possible transmissions are controlled by the NAV in the CTS, see Figure 4.

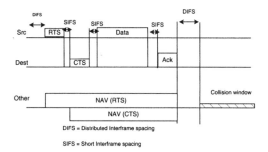

Figure 4. RTS/CTS scheme and Network Allocation Vector

3.2 An optimized IEEE 802.11 MAC model

3.2.1 Generalities

In the following we will describe an optimized model for the IEEE 802.11 MAC scheme. This optimization will try to maximize the speed of the simulation and will sometimes lead to a slight simplification or approximation in the modeling. Of course, the speed of a simulation is proportional to the number of events generated in a simulation. Therefore, in the following, the number of simulation events used will be of prime importance. We will try, as far as possible, to minimize this figure, which sometimes leads to slight approximations. All these approximations will be justified and discussed.

3.2.2 Simulation of collision

In a CSMA system a collision can occur in the two following situations:

– two transmissions start approximately at the same time so the transmission of the other node has not yet been sensed due to propagation delay and to electronic detection delays;[1] this situation is usually called a collision.

– two nodes are hidden from each other and so carrier sensing does not operate between these nodes; a collision can occur at any time during the life time of the packet. This situation is usually called "hidden node collision" or "late" collision.

Actually, simulating collision is not difficult since the simulator will rule the transmission according to the carrier sense indication exactly as in a real network. However, the simulation of the collision generally uses a usual technique often called the "collision window" technique. To take into account propagation and electronic detection delay we will need to distinguish the starting transmission time at the source and the time when the reception starts in receivers. This will require creating two different events: one event at the transmitter and one event within all the potential receivers to indicate the start of effective reception of the transmission. This latter event is called the "start of carrier". The time interval between these two events is often called the collision window, see Figure 5. Indeed, it is in this window that starting an alien transmission will create a collision.

   Of course a possible optimization could be not to send the event "start of carrier" to distant nodes, as this event has no effect on distant nodes. But implementing this optimization is not easy especially in a network with mobile nodes. Therefore this optimization will not be used in the model presented.

---

[1] In networks of few hundred meters of average size, the electronic detection delays lead to the predominant factor.

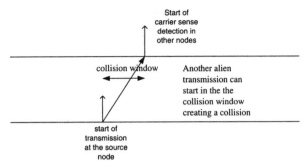

Figure 5. *Collision window*

3.2.3 Simulation of the back-off

We have already seen that the IEEE 802.11 back-off is slightly peculiar. Indeed, it requires that during this back-off period the nodes constantly monitor the channel to sense the carrier activity. In the standard the back-off is a multiple number of slots (actually called collision slots). A direct monitoring of the channel leads to scheduling an event for every collision slot. Of course, this approach is not efficient since it leads to the creation of a lot of unnecessary events. This can be avoided if we are able to destroy events. In this case all nodes will schedule its "end of back-off" event. When the closest "end of back-off" comes to be treated, the carrier sense reception by the nodes currently in back-off leads these nodes to stop their back-off decrementation and to register their remaining back-off duration. The carrier sense returning to idle allows these nodes to reschedule their end of back-off event with the remaining back-off duration (see Figure 6). This simulation technique allows an exact description of the back-off scheme while it significantly reduces the number of simulation events.

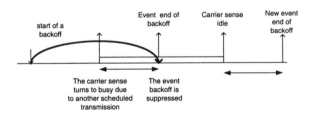

Figure 6. *Simulation of the CSMA/CA back-off*

3.2.4 Simulation of the acknowledgement

A transmission without acknowledgement requires two events:

    – the end of a back-off when the node starts transmitting;
    – the end of a transmission.

A transmission with acknowledgement requires four events:

– start of the transmission at the end of a back-off;
– end of a transmission;
– start of the acknowledgement transmission;
– end of the acknowledgement transmission.

Therefore, roughly speaking, the acknowledgement doubles the required number of events for a transmission. There is a means to simulate the acknowledgement with only two events as shown in Figure 7. The frame duration is increased by the acknowledgement duration. Here, there is of course a slight approximation since the carrier activity is slightly different, however, this effect is very limited since the acknowledgement duration is small. There is another effect in the possible collision of the acknowledgement packet. But for the same reason this event is very unlikely and is necessarily created by a hidden node. On the other hand, the simulation model will take into account a collision on the destination during the acknowledgement. This collision will also be produced by a hidden node. The real scheme and the simulation model will therefore show very similar performances.

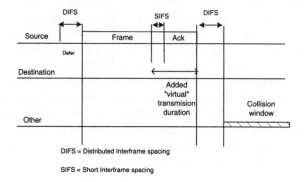

Figure 7. *Simulation model for the acknowledgement*

There is another important "trick" that is worth mentioning concerning the simulation of the acknowledgement. With the proposed model we do not have the "real acknowledgement" to detect collision. We therefore need to "visit" the reception node just before ending the transmission to check whether the transmission is a success or a collision. That is why, in the simulation model, we have scheduled the end of the transmission e bits before the end of the transmission; e must be small and we have taken $e = 1$, see Figure 8.

3.2.5 Simulation of the RTS/CTS

In the RTS/CTS scheme we have the following events:

– start of the RTS transmission at the end of a back-off;
– end of RTS transmission;

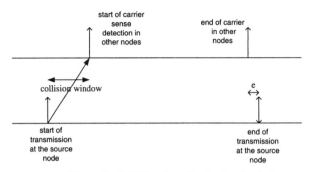

Figure 8. *Collision detection in the simulation*

– start of the CTS transmission;
– end of the CTS transmission;
– start of the frame transmission;
– end of transmission;
– start of the transmission of the acknowledgement;
– end of the transmission of the acknowledgement.

Thus a complete transmission process costs eight events. Actually as seen in the previous subsection it is possible to ignore the two events related to the transmission of the acknowledgement. There remain six events for the complete transmission process and we can reduce this number to three.

Let us consider the following events:

– start of the RTS transmission;
– end of the RTS transmission, depending on the correct reception of the RTS by the destination, this event is or is not followed by the "full" transmission of the frame (CTS + frame + Ack);
– the end of the transmission (only applicable if a transmission has started!).

The model is described in Figure 9. When there is no collision on the RTS, the CTS, the transmission of the frame and the acknowledgement are concatenated. If there is a collision for the RTS the transmission is stopped. In the simulation code at the end of the RTS transmission we must visit the destination to see if the RTS has been correctly received. If so, we must inform the neighbors of the receiver of the forthcoming transmission duration to simulate the NAV effect. We can see that the main different between the real IEEE 802.11 operation and the proposed simulation model lies in the fact the CTS transmission is, in the simulation model, supposed to be transmitted by the source whereas in reality it is transmitted by the destination. Thus, in principle, the simulation model does take into account the collision on the CTS. But with correct operation this situation is not possible or very unlikely. A potential collider on the CTS must be within reach of the sender. But in such a case the RTS should have been received by this collider and the collider should be informed of the forthcoming transmission activity.

We can implement this model by using the event "end of RTS". We have to visit first the destination to check if the RTS has been correctly received. If so, the transmission is continued; if not, the transmission is stopped. We have to visit the event "end of RTS" at the source to apply the transmission decision. Therefore it is sensible to schedule the end of the RTS 1 bit earlier at the destination node than at the source node.

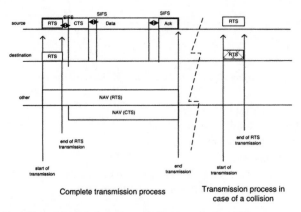

Figure 9. *Simulation model for the acknowledgement*

3.2.6 Reception engine

The reception engine uses a three-state automaton. These three states are:

– "idle", there is no carrier detected;
– "carrier", a carrier is detected but a comprehensive packet can not be decoded;
– "reception", a comprehensive packet is received.

The transitions between these states are simple and can only occur at the beginning of a transmission; this helps to avoid unnecessary events.

3.2.7 The OPNET simulation tool

The OPNET simulation tool is one of the most widespread tools. This tool provides the following services:

– a scheduler;
– an easy way to code state automatons;
– a very powerful graphic interface to present simulation results.

OPNET also contains a lot of codes dedicated to simulating radio links, access protocols and various protocol layers. We have tried to using these codes but they resulted in simulations which were long in duration. Since the speed of our simulation was of prime importance to us, we decided not to use any of these facilities.

4. Simulation results

4.1 IEEE 802.11 figures

In Figure 10, we show the structure of an IEEE 802.11DS packet at the physical layer. The physical layer encapsulation has a duration of 192 μs. In Table 1 we give the duration of various IEEE 802.11 attributes. These durations allow us to compute the overhead which, in duration, slightly depends on the bandwidth due to the 34 octets of the MAC overhead. The IEEE 802.11 DS overheads are given in Table 2. The following simulation results will take these overheads into account. The offered load will therefore be the payload i.e. the load of useful data. We will be able to offer a precise performance evaluation of the IEEE 802.11 DS standard.

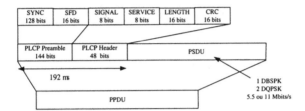

Figure 10. *Structure of an IEEE 802.11 packet*

Table 1. *Main IEEE 802.11 figures*

IEEE attribute	Duration in μs
DIFS	50
slot time	20
SIFS	10
Phy overhead	192
MAC overhead	34 octets
Acknowledgement	304 at 1 Mbps

Table 2. *IEEE 802.11 overheads*

Air rate	Duration in μs
1 Mbps	778
2 Mbps	642
5.5 Mbps	555
11 Mbps	530

4.2 IEEE 802.11 performance analysis with various traffic scenarios

4.2.1 Simple scenarios

All the nodes are within range. We suppose that we have 2, 5, 10, 15 or 20 nodes with various data rates: 1 Mbps, 2 Mbps, 5.5 Mbps and 11 Mbps. We generate a traffic with large packets: 12000 bits and we test the network with a load ranging from 10% to 100% of the channel capacity by steps of 10%. We do not use the RTS/CTS option. The result of this simulation allows us to find out the maximum channel capacity of the IEEE 802.11 standard. The results are given in Figures 11, 12, 13. As expected, we can see that the channel throughput decreases as the number of nodes increases. That is a general result of the CSMA scheme. We can also see that the normalized channel throughput decreases as data transmission rate increases. This phenomenon can be explained by the fixed overhead in the frame and also because the normalized size of collision window by the frame transmission duration increases (see the analytical model provided below).

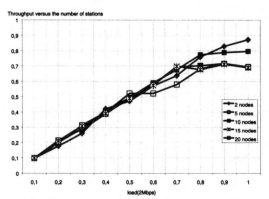

Figure 11. Results of IEEE 802.11 DS at 2 Mbits/s

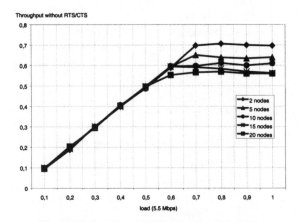

Figure 12. Results of IEEE 802.11 at 5.5 Mbits/s

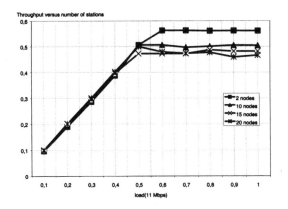

Figure 13. Results of IEEE 802.11 at 11 Mbits/s

To investigate the effect of the RTS/CTS option on the achievable throughput the same simulations as those above were carried out with this option. Figure 14 gives the simulation results at 1 Mbits/s with 20 nodes with and without RTS/CTS. We then investigate at 11 Mbits/s with and without RTS/CTS; the results are given in Figure 15.

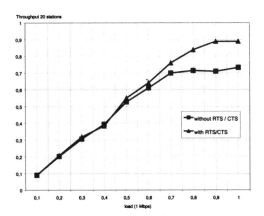

Figure 14. Results of IEEE 802.11 with the RTS/CTS and 20 stations at 1 Mbits/s

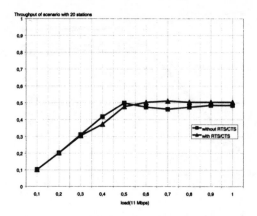

Figure 15. Results of IEEE 802.11 with the RTS/CTS and 20 stations at 11 Mbits/s

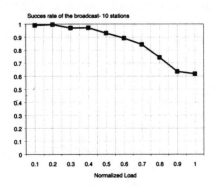

Figure 16. Percentage of received broadcast with 10 stations sending 2 kbits packets

4.2.2 Scenarios with broadcast

We next focus on the broadcast issue. We have studied the scenario where all the nodes send a broadcast traffic and we investigated the success rate. We present the results with 10 stations sending broadcast packets of 2000 bits, as shown in Figure 16. The simulation results show that the collision rate is more than 10% for a load greater than 50% of the channel capacity. This bad performance for broadcast traffic is an issue for the IEEE 802.11 standard.

We also studied the scenario where all the nodes send a broadcast traffic and we investigated the collision rate. We present the results with 10 stations sending broadcast packets of 2000 bits. The simulations show that the collision rate for the broadcast traffic can be quite high. Actually, in all of the scenarios mentioned above we experienced collision rates greater than 15.

4.2.3 Scenario with hidden nodes

We investigated the utilization of RTS/CTS in the case of hidden nodes. In the following scenario four stations are placed at the corner of a square 40 m × 40 m. Two nodes, node 1 and 2, are not within carrier sense reach; for instance there is a steel obstacle between the two nodes. Therefore node 1 and 2 are hidden from each other and can create hidden collisions. The transmission rate of this scenario is 2 Mbits/s, and is illustrated in Figure 17. The simulation results show that at high load the RTS/CTS scheme can save around 10% of the channel capacity. We could expect this result since the NAV mechanism can save bandwidth.

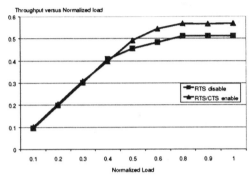

Figure 17. Channel throughput with 4 stations with hidden nodes at 2 Mbits/s

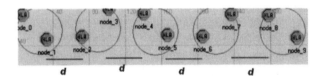

Figure 18. Scenario of spatial reuse with ten stations in 5 groups of 2 nodes

4.2.4 Scenarios with spatial reuse

In the following sections, we will study scenarios involving spatial reuse. The scenarios proposed use 10 stations which are divided into 5 groups of two stations. This scenario configuration is shown in Figure 18. The 5 groups of two stations are separated by a distance of D meters. We study the total throughput of the network versus the distance D and we use the RTS/CTS. The simulation results are given in Figure 19. Of course as D increases the total throughput increases since the 5 groups tend to be more independent. The results given in Figure 19 are coherent with the results in Figure 15. As a matter of fact when the 5 groups are independent we find a throughput five times greater than the throughput of a single group.

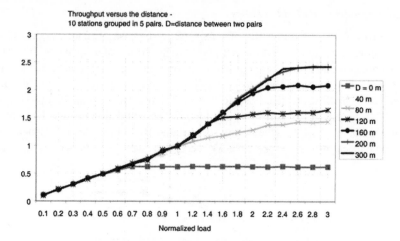

Figure 19. Channel throughput with ten stations in 5 groups of 2 nodes versus the distance between the groups of nodes

4.3 Comparison of simulation results with a simple analytical model

In this part our aim is to compare the results of a simple analytical model with the simulation results. We use a classical model of CSMA which can be found in [7]. The assumptions of this model are the following:

- the packet has a duration of 1;
- β is the duration of the collision window;
- λ is the Poisson arrival rate;
- the number of stations in back-off is n;
- q_r is the retransmission probability.

With these assumptions it is possible to study the variation in the number of waiting stations. Let us denote by D_n the drift of this number of waiting stations. D_n can be computed as the difference between the expectation of arriving stations in back-off minus the expectation of a successful transmission. It can be shown that

$$D_n = \lambda\,(\beta + 1 + e^{-g(n)}) - g(n)e^{-g(n)}$$

where

$$g(n) = \lambda\beta + q_r n$$

The protocol is stable if the drift is negative, in which case we have:

$$\lambda < \frac{g(n)e^{-g(n)}}{\beta + 1 + e^{-g(n)}}$$

It can be easily derived that maximum channel efficiency is obtained with $g(n) = \sqrt{2\beta}$. The obtained valued of the channel efficiency is then:

$$g(n) = \frac{1}{1 + \sqrt{2\beta}}.$$

To use this simple model we have now to compute the value of β corresponding to our simulated scenarios. β measures the value of the collision window given that the transmission duration is 1. Therefore at 11 *Mbits/s* with a collision slot of 20 μs with a packet length of 12000 bits and taking into account the propagation delay,

$$\beta = \frac{20}{530 + \frac{12000}{11}} = 0.01233$$

We can build Table 3 which gives the maximum channel efficiency versus the data rate.

Table 3. Maximum channel capacity

Air rate	$\frac{1}{1 + \sqrt{2\beta}}$
1 Mbps	0.968
2 Mbps	0.927
5.5 Mbps	0.892
11 Mbps	0.864

Table 4 shows the results of computing the overhead factor for 12000 bits packets i.e. the fixed overhead due to the IEEE 802.11 overhead.

Table 4. Transmission overhead

Air rate	overhead factor
1 Mbps	0.939
2 Mbps	0.903
5.5 Mbps	0.797
11 Mbps	0.673

In a simple approximation the channel throughput that we have measured is equal to the product of the maximum channel efficiency by the overhead factor. We obtain the following results given in Table 5.

Table 5. *Maximum normalized throughput*

Air rate	maximum normalized throughput
1 Mbps	0.91
2 Mbps	0.84
5.5 Mbps	0.710
11 Mbps	0.581

Since the assumptions of the model [7] assume a perfect stabilization of the retransmission protocol the above results give upper bounds for the real figures. We can easily check that on the simulation results.

5. Conclusion

In this paper we have proposed a simple simulation model for wireless LANs. At the physical layer, the only important assumption is the linearity of the model; however it can accept all kinds of propagation model with decaying, fading, interference etc. At the MAC layer we have proposed a simple model for the CSMA/CA scheme including the acknowledgement and the RTS/CTS (if activated). This model is optimized to provide a fast simulation tool of the IEEE 802.11 MAC. We have proposed the simulation results of various scenarios using the exact IEEE 802.11 overhead. We have given thus a detailed performance analysis of the IEEE 802.11 protocol with various data rates. We have most particularly studied the maximum channel throughput, the performance of the protocol for broadcast traffic, the effect of the RTS/CTS, channel reuse. In future work, we will use this model to evaluate routing protocols. We also plan to use this simulation model to evaluate various QoS proposals for the IEEE 802.11 standard.

REFERENCES

[1] ETSI STC-RES 10 Committee, HIPERLAN Functional Specifications, draft standard ETS 300–652, 1995.
[2] IEEE 802.11 Standard. Wireless LAN Medium Access Control (MAC) and Physical Layer (PHY) Specifications. June 1997.
[3] ANSI/IEEE Std 802.3, 2000 Edition. Information Technology–Local and Metropolitan Area Networks–Part 3: Carrier Sense Multiple Access with Collision Detection (CSMA/CD) Access Method and Physical Layer Specifications.
[4] METCALF, R. M., BOGGS, D. R., 1976. Ethernet: Distributed Packet Switching for Local Computer Networks. *Comm ACN*. 395–404.
[5] 8802-5: 1998 (ISO/IEC) [ANSI/IEEE 802.5, 1998 Edition] Information Technology–Telecommunications and Information Exchange Between Systems–Local and Metropolitan Area Networks–Specific Requirements–Part 5: Token Ring Access Method and Physical Layer Specifications.

[6] MATHYS, P., FLAJOLET, P., Q-ary Collision Resolution Algorithms in Random-access Systems with or Blocked Channel Access, in *IEEE Trans. on Information Theory*, vol. IT-31, p. 217–243, 1985.

[7] BERTSEKAS, D., GALLAGER, R., Data Networks, Prentice Hall, 1988.

Chapter 4

Mobility management in a hybrid radio system

Matthias Frank and Wolfgang Hansmann
University of Bonn, Germany

Tomas Göransson and Ola Johansson
Ericsson Mobile Data Design, Sweden

Thorsten Lohmar and Ralf Tönjes
Ericsson Eurolab Deutschland GmbH, Germany

Toni Paila and Lin Xu
Nokia Research Center, Finland

1. Introduction

The demand for access to mobile multimedia services at any time and anywhere has driven the development of 3G mobile communication systems. The upcoming 3G could be labeled as mobile multimedia as it enables a multitude of services and clearly dominates future network evolution activities on all levels, building on two key technologies: the Internet and digital cellular mobile radio systems.

However, the demand for cost efficient provision of mobile multimedia services is faced with the reality of scarce radio resources. The requirement of spectrum efficiency has driven the development of various digital radio systems that have been optimized for mobile communication, wireless access, or broadcast, respectively (e.g.: GSM – *Global System for Mobile Communication*, GPRS – *General Packet Radio Service*, UMTS – *Universal Mobile Telecommunication System*, WLAN – *Wireless LAN*, DAB – *Digital Audio Broadcast*, DVB-T – *Digital Video Broadcast Terrestrial*). In the future, wireless systems will contribute to the integration of different access technologies and their convergence. Future wireless systems will feature a user-friendly communication platform that provides cost effective bandwidth to access personalized services in a seamless manner.

The European research project *DRiVE* (Dynamic Radio for IP Services in Vehicular Environments [DRi01]) aims at enabling spectrum-efficient high-quality wireless IP in a heterogeneous multi-radio environment to deliver in-vehicular multimedia services. DRiVE addresses this objective on three system levels: The project investigates methods for the coexistence of different radio systems in a common frequency range with dynamic spectrum allocation.

DRiVE develops an IPv6-based mobile infrastructure that ensures the optimized inter-working of different radio systems (GSM, GPRS, UMTS, DAB, DVB-T) utilizing new dynamic spectrum allocation schemes and new traffic control mechanisms. Furthermore, the project designs location dependent services that adapt to the varying conditions of the underlying multi-radio environment.

DRiVE addresses the key issue of spectrum efficient resource utilization by two functions: *dynamic spectrum allocation* (DSA) and *traffic control* (TC). The idea behind DSA is that radio spectrum, which is allocated to specific radio access systems, may not be utilized fully at all times and in all regions. Hence this spectrum could be used by other systems for services not provided by the dedicated system. By introducing DSA in the radio access systems, spectra could be moved between the systems to optimize the overall usage in an area. A dynamic hybrid radio system requires that the system announces which access technology is available in which frequency range. A generic means to provide this information is a (logical) *common co-ordination channel* (CCC). DSA is out of the scope of this article and the interested reader is referred to [Lea01].

Traffic control co-ordinates the distribution of a mobile node's traffic on the available access systems. The distribution depends on different parameters such as: the user preferences (preferred access system, cost preferences), terminal related information (terminal capabilities, location, and signal reception quality, etc.), the traffic parameters (QoS requirements), the status of the network (load, available capacity) and service requested. Both host and network can control the distribution of traffic. This article proposes a mechanism for *host controlled flow based forwarding* that allows a mobile terminal to select the most appropriate access technologies for different types of traffic. The described IPv6 based mobile multi-access infrastructure enables the co-operation of existing access networks for spectrum efficient provision of mobile multimedia services, supporting in particular also asymmetric and interactive services.

[Zha98] has already presented some ideas for a flexible support of mobility in allowing a mobile host access to several network interfaces in parallel to select the most appropriate one for both upstream and downstream flows. However, in the context of IPv6 and micro-mobility support (cf. [Cam00]), the DRiVE approach eases the burden for the correspondent nodes in the Internet by providing a central control node within a DRiVE backbone network (cf. Sections 4 and 6 for more details).

The article is structured as follows: Section 2 briefly motivates the key aspects of a hybrid multi-radio architecture and Section 3 describes the multi-access architecture and its basic functionality. Section 4 highlights the main challenges of the multi-access scenario for mobility support and the following section presents a solution for mobility management in DRiVE making use of current IETF approaches with DRiVE specific enhancements. Section 6 discusses the traffic control functionality needed to distinguish between downstream flows with different *Quality of Service* (QoS) requirements, resulting in a selection of different radio access systems. Finally, Section 7 concludes this contribution.

2. Scenario

Traditional Internet applications like web browsing, e-mail, push services are in many cases asymmetrical in their nature, generating more downlink than uplink traffic. For a better support of these applications more downlink capacity than uplink capacity should be provided. It is also possible to use different radio access systems for uplink and downlink respectively to further enhance the spectra usage or to provide additional downlinks in a different frequency range. For the sake of simplicity, we associate a radio access system with each frequency range. This leads to the envisioned scenario, where a mobile user is able to communicate via several radio access systems at the same time. There are several solutions that could be employed to distribute the traffic load onto several bearers.

The DRiVE approach is to separate multimedia sessions and to transfer multimedia flows via different access systems. In this article, we assume that each system is optimized to specific types of services. For instance, an access system *A* may be designed to reserve some capacity to assure a certain quality of service to a traffic flow, but only at a rather low data rate. Another access system *B* operating in a different frequency range may provide some additional capacity and higher available bitrates, but only as a best effort service. Multimedia services usually consist of several traffic flows with different traffic characteristics. To utilize all available access systems in the most efficient way the traffic flows of the user are distributed to the most appropriate access system(s).

3. Multi-access architecture

Multi-access is the capability to connect a terminal to several network attachment points of different technologies simultaneously for obtaining access to the same application services. In addition, each access system may provide further different application services. There can be simultaneous connections to different access systems, or connections to only one access system at a time.

In DRiVE, we have selected an overlay design for the multi-access architecture (cf. Figure 1). The main function of the network infrastructure is to provide mobile users communication services over heterogeneous access systems and to manage user mobility. We can divide the entire system architecture into two logically separate parts: an access system independent and an access system specific part. Here, by the term access system we denote a completely stand-alone system comprising both the radio network part as well as the core network part.

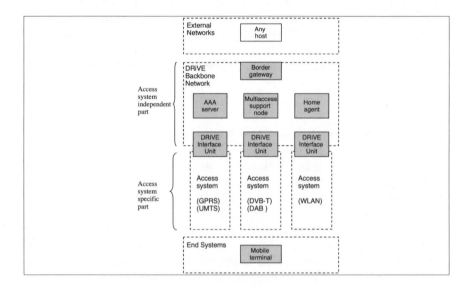

Figure 1. *DRiVE network architecture*

Multiple individual access systems form a logical part of the system. By individual we mean two aspects: the access systems can be administrated by different operators and the access systems do not need to know anything about each other. We have designed the architecture in such a way to handle access systems as black boxes. The minimum requirement is that an access system has to enable delivery of IP datagrams to and from a *DRiVE mobile terminal* (DMT). All the other functions including mobility management and *authentication, authorization and accounting* (AAA) are handled by the access system independent part. Figure 2 gives an example of a potential operator clustering case.

A wide variety of potential access technologies also means a broad range of characteristics among the systems. Communication between a mobile user and the network can be uni-directional (broadcast systems) or bi-directional (cellular and hot spot systems). Broadcast systems have large cell sizes as well as high but shared uni-directional (downlink) bandwidth. Additionally, they allow relatively fast terminal mobility. Cellular systems offer bi-directional access with high tolerance for terminal mobility. However the bandwidth is limited. Hot spot systems, such as WLAN, offer varying bandwidth as their radio techniques are based on a shared channel. Thus, the data rate varies depending on how many users access the same access point. Movement can cause a lot of variation in the

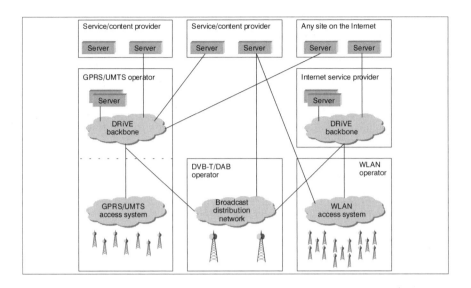

Figure 2. Example of operator clustering

performance, too, as cell sizes are small and the specification of handover is not very sophisticated. Thus, the allowed mobility is close to stationary.

The access system independent part in the DRiVE architecture is the DRiVE backbone network. It is effectively an IPv6 capable network enhanced by special system entities described later in this article. Figure 1 shows how these system entities are placed on the architecture. We want to especially emphasize two logical system entities. Firstly, the *DRiVE interface unit* (DIU) makes it possible to interface virtually any IP capable access system with the backbone. Secondly, the *multi-access support node* (MSN) implements the traffic distribution. Also, it is worth noting that all the access technology independent functions are implemented on IP level. Thus, the DRiVE backbone can be seen as a part of an ISP core or even a UMTS core network, for example.

4. Inter-system mobility

Each access system in a DRiVE system enables a DMT to connect to the IPv6 based DRiVE backbone network. Different radio access systems are optimized for different type of services and have different cost vs. bandwidth value. The DRiVE system enables a mobile terminal to select the most appropriate bearer for each of its services and to route traffic of different services via different radio access systems.

Since DRiVE provides a wireless environment that allows mobile terminals to send and receive IP packets while in motion, a DMT may be out of the coverage area of one access system during an on-going session. To enable continuity of a session, the DMT needs to perform an inter-system handover and route its on-going session via another access system (cf. Section 6).

Furthermore, the traffic control function of the DRiVE system enables a DMT to select an access system that is capable of providing QoS set by the application. However, as the traffic load changes, the QoS provided by one access system also varies. An access system may be able to provide a certain QoS during some time periods, but may be unable to provide the same QoS during other time periods. In order to maintain the desired QoS, a mobile terminal may need to reselect the access system during a session and perform an inter-system handover.

Figure 3 depicts an example of coverage of a few heterogeneous access systems. In area 1 the DMT is in coverage of GPRS and a broadcast system like DAB or DVB, which provides high bandwidth but is only available for downstream communication. In area 2, UMTS and WLAN is available, both allowing for uplink and downlink communication. On the move between area 1 and area 2, the DMT has to switch between DxB, GPRS, UMTS and WLAN and also has to pass a region where only UMTS is available before getting to the WLAN area. Moreover, the fact that a broadcast system like DAB or DVB is only available for downstream communication imposes another challenge to the mobility management.

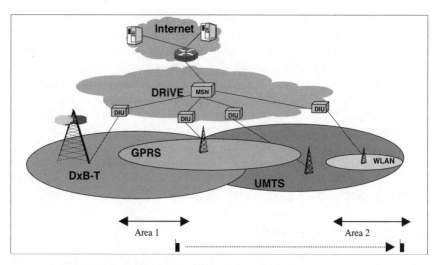

Figure 3. *Heterogeneous access system coverage in different areas*

For the above two reasons, the DRiVE system needs to provide support for flexible inter-system mobility. Moreover, the inter-system mobility support provided by the DRiVE network should enable traffic of a single session to be redirected.

Currently, none of the existing mobility support schemes satisfy these requirements. If we use the existing schemes, for example Mobile IP, all the IP packets heading to a single DMT will be routed via a single access system. Thus, the DRiVE project needs to extend the existing scheme to fulfil the special mobility requirement imposed by the DRiVE system.

5. Mobility management in DRiVE

A common design goal of cellular access systems such as GSM or UMTS is to support mobility of user devices. These access systems implement their mobility management which cares for proper delivery of packets to the current position of a mobile device. This is done transparently to a communicating peer in a fixed network. However, other access systems that have been designed for broadcasting (e.g. DAB and DVB-T) or local wireless access (e.g. Bluetooth) do not provide mobility support or have only limited layer 2 based mobility support (e.g. WLAN, IEEE 802.11) within the own access system.

To integrate these different types of access systems with their different capabilities to handle mobility into a hybrid radio system, an access system independent mobility management is needed. As, apart from using radio for transmission, the common feature of all access systems is the capability to transport IPv6 packets, mobility management in DRiVE can only be orientated to a network layer based solution.

Mobile IPv6 (cf. [Joh01]) is currently in its last stage towards completion by the IETF *mobileip* working group. It is an extension to the specification of IPv6 to permit mobile terminals to be reachable by a single address (their *home address*) independent of their current attachment to a network. In addition to its home address, a mobile terminal receives a *care-of-address* from its current network. A *Home Agent* is a special router in the mobile terminal's home network: it intercepts packets sent from fixed nodes in the Internet (*correspondent nodes*) directed to a roaming mobile terminal and actively forward it to the mobile terminal's current *care-of-address*.

The specification of Mobile IPv4 (cf. [Per96]) mandated the use of a Home Agent to forward packets to a mobile terminal. A mobile terminal usually sends packets directly to a correspondent node, while in the reverse direction packets had to traverse the Home Agent. Therefore, the basic Mobile IPv4 operation is often referred to as *triangular routing*. If a mobile terminal's home network is far from its current network, it is obvious that triangular routing introduces significant routing overhead. To overcome this drawback, Mobile IPv6 permits direct communication between a mobile terminal and a correspondent node. This is achieved by a *binding cache* maintained at each correspondent node. A binding cache maintains home address to care-of address mappings. If a binding cache entry exists for a mobile terminal, a correspondent host sends packets directed to this mobile terminal to its care-of-address instead of its home address. A mobile terminal properly sets up its binding cache entry on a correspondent node by sending a *Binding Update* message.

Despite the advantage of route optimization, Mobile IPv6 suffers a drawback during handover. Each time a mobile terminal changes its foreign network, all of its corresponding nodes have to be notified by a Binding Update message (also cf. [Cam00]). Thus, packets directed to a mobile terminal will arrive at its new position at least one round trip time after a binding update was sent. Depending on the application (e.g. voice over IP) traffic disruption for one round trip time may lead to a perceptible decrease of service quality. A solution to reduce the signaling latency during handover is *Hierarchical Mobile IP* (HMIP, cf. [Sol01], [Cas00]), which is currently discussed by the IETF mobileip working group as an enhancement to Mobile IPv6. HMIP proposes a new network entity, a *Mobility Anchor Point* (MAP) which is located near a mobile terminal's foreign network. Instead of its real care-of-address, a mobile terminal registers a *regional-care-of-address* at its correspondent node. The regional care-of address is obtained from a mobile terminal's MAP. Thus, traffic directed to the mobile terminal is routed towards the MAP, which forwards it to the mobile terminal's current care-of-address. If a mobile terminal performs a handover, it just registers its new care-of-address at its MAP. Correspondent nodes are unaffected.

Obviously, HMIP greatly reduces the amount of Binding Update messages to be sent after a handover. One could argue that the MAP reintroduces triangular routing that Mobile IPv6 wanted to eliminate. However, the routing overhead introduced by the MAP is considerably lower than triangular routing via the Home Agent as the MAP is located much closer to the mobile terminal than its Home Agent. Another advantage is handover efficiency: as the MAP is near the mobile terminal's location, handover signaling is faster than with plain Mobile IPv6. Real-time services to be delivered to mobile terminals require low service disruption, which HMIP greatly contributes to.

The DRiVE architecture implements a MAP in the MSN and the DIU service entities (cf. Figure 1). It uses an enhanced version of HMIP, which is enhanced for host controlled forwarding (see next section). Figure 4 depicts the use of HMIP in a scenario with a WLAN network with large coverage having several access routers (AR). These ARs are connected separately to the DRiVE core network and are executing the DRiVE Interface Unit (DIU) system entity functionality. The MSN of the DRiVE core network is acting as the Mobility Anchor Point of HMIP. After the mobile terminal has moved to the coverage area of another WLAN AR (Figures (a) to (b)), the mobility binding update has to be performed. In step (c) the Binding Update is performed between the DMT and the MAP within the DRiVE MSN. In Figure (d) upstream and downstream traffic from correspondent node(s) are routed to and from the new location of the DMT and the corresponding new WLAN AR.

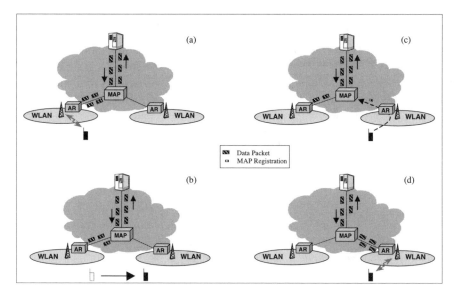

Figure 4. *Applying HMIP in DRiVE, general example*

6. Host controlled flow based forwarding

Usually a DMT is in the transmission range of several different access system technologies. However, coverage from high-rate systems like wireless LANs with "hot spot" coverage may be lost from time to time as the DMT moves. Most access systems considered in the DRiVE project have disparate characteristics regarding e.g. bandwidth and delay, which means they are suitable for different types of traffic. At the same time a DMT can be connected to several of these access systems, provided it has the proper receiver and subscription. Consequently, it could be beneficial for the DMT to be able to choose to use one access system for one type of traffic and another for another type of traffic. As an example, consider the following scenario: a DMT is connected to a GPRS system, which is the default path for best effort traffic like e-mail. A more bandwidth demanding service like streaming audio is requested. This will result in a moderate perceived quality, but as the DMT moves it may discover that it has reached the coverage area of a WLAN system. It then would like to move the streaming audio flow to the WLAN system to enhance the perceived quality. The choice could be made considering type of traffic as well as the cost of using an access system. Decisions could be made either manually or automatically based on a predefined policy.

Naturally, the DMT can choose to send a regular MIPv6 binding update to its corresponding node containing the care-of-address it uses in the preferred access system. However this means that the benefits gained from the hierarchical mobility management are lost (cf. previous section). Instead, the proposed course of action in DRiVE is to let the MSN, which is the central point in the DRiVE network, be the point where downlink traffic is directed according to the wishes of the DMT (cf. Figure 5). This is accomplished through an extension of the hierarchical mobility management scheme (also cf. the previous section). The binding cache of the MSN is expanded to also store identifiers of a particular flow or connection, which are mapped to different access systems corresponding to different care-of-addresses. This of course requires new binding update options to carry the flow identifiers in addition to the associated care-of-addresses. Throughout this section the term binding update should be interpreted as a binding update including the new extension.

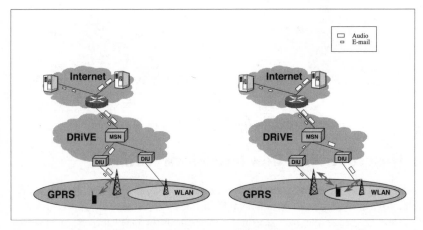

Figure 5. *(left) All traffic directed through the GPRS system; (right) the MSN directs the audio flow through the WLAN system*

As an identifier for a flow (or a connection) one field or a combination of several fields within the IPv6 header and the transport protocol header can be used. Suitable fields within the IPv6 header are the flow label, the source address and the next header field (indicating the transport protocol). The traffic class field is less advisable to use since routers along the path are allowed to alter its value as stated by the general requirements in the IPv6 specifications (cf. [Dee98]). The source and destination port fields of the transport protocol header are eligible in a layer-crossing multi-field combination. Two plausible combinations of the abovementioned fields as flow identifiers are 1) the source address field and flow label, or 2) the source address, the protocol identifier and the source and destination port. We propose that only one or maybe two different field combinations will be used in order to identify different flows at the MSN. This is for the sake of keeping the processing burden low.

If a DMT's binding cache entry has more than one care-of-address mapped to its home address, the MSN will compare incoming packets with the identifier of each care-of-address. If a match is produced, the packet will be forwarded to the care-of-address corresponding to the identifier that matched the current packet. When the MSN fails to find a match, packets are forwarded to the primary care-of-address of the DMT.

There are two alternatives of when to send a binding update for the cause of (re-) directing a flow provided that the DMT is covered by more than one access system at a time. If the DMT is aware of the requirements of a flow, it can send the binding update prior to requesting the service. Otherwise, if the DMT is unaware of a flow's requirements it can start the session and (by some mechanism that is outside the scope of this article) discover that the available bandwidth is insufficient to provide the desired level of quality and then send the binding update. In the first case the forwarding path is determined before the session is initiated and in the second case traffic is redirected during the ongoing connection.

To continue the previous example, consider a DMT receiving traffic through its default access system (GPRS) while moving into the range of a WLAN access system also connected to the DRiVE backbone. The extended mobility management functionality decides to move the flow to the WLAN system. The decision to redirect a flow is based on a list of available access systems and a measure of the quality of service. The DMT sends a binding update containing the care-of-address it has obtained from the WLAN system together with e.g. the flow label (which it has learned from packets arriving through the GPRS system) and the source address of the server providing the audio service, as an identifier of the flow. The MSN updates its binding cache with the new entry. All subsequent packets addressed to the DMT matching the source address and the flow label provided in the before mentioned binding update will be forwarded over the WLAN by the MSN. If the DMT moves out of the coverage of the WLAN while still keeping the flow active a new binding update has to be sent to redirect the flow back to the GPRS system. Note that in this example only the streaming audio flow is redirected. All other traffic is forwarded through the default system unless other binding updates were sent.

7. Conclusions

In this article we have presented the DRiVE system with a focus on two of several challenging issues of the DRiVE concept: mobility support and host controlled flow based forwarding of downstream packets via different radio access systems. Both issues play an important role in a scenario with dynamically changing availability of several radio access networks with different service capabilities.

An important prerequisite on using several independent radio access networks to provide multimedia services is the operation of an access system independent network part which allows for a hybrid usage of the radio access systems. The DRiVE network architecture defines the DRiVE backbone, which fulfils central functions like user management, authentication, authorization, accounting and others. The architecture is flexible enough to allow for different operator roles for cellular network operators, DRiVE and other services providers, broadcast operators and Internet Service Providers.

One of the main challenges is global mobility support of mobile users moving and possibly leaving or entering the coverage area of specific radio access systems. In this case, an inter-system handover may be useful to change to an access system better fulfilling the service requirements. Or, the handover may be necessary because the former best suitable access system is getting out of radio coverage. The DRiVE traffic control function allows for the selection of the appropriate radio access system. The mobility of the DMT is supported in a combination of Mobile IPv6 and Hierarchical Mobile IP: On the longer transmission leg between a correspondent node far outside in the Internet world and the Multi-access Support Node MSN of the DRiVE backbone, the MIPv6 approach of binding caches helps to prevent the worrisome triangular routing via the Home Agent. On the shorter transmission leg between the MSN and the mobile terminal the Hierarchical Mobile IP concept is applied.

The availability of several radio access networks introduces the opportunity to use these access networks in parallel to fulfill different QoS requirements of different parallel downstream data flows. This results in the challenge of a per-flow decision of where to route downstream IP datagrams. The MSN within the DRiVE backbone network is the splitting point of different downstream data flows. The mobile terminal may register several care-of-addresses of different radio access systems with the MSN in parallel, some being specific for an individual flow considering its QoS requirements and the capabilities of a specific radio access system. As with hierarchical mobility support, host controlled forwarding is limited to the shorter transmission leg between MSN and the mobile terminal.

Further work will improve mobility support, in particular for inter-system handover, host and network controlled flow based forwarding, and Quality of Service support. The project will contribute to IETF activities in the area of mobility support and Hierarchical Mobile IP. DRiVE has demonstrated its key concepts at the IFA (Internationale Funkausstellung Berlin – International Radio Communication Fair, cf. [IFA01] and [Ald01]) and will perform user trials to validate the concepts and technological advances. Results of the demonstration outcomes will be published in separate publications.

Acknowledgements

This work has been performed in the framework of the IST project IST-1999-12515 DRiVE, which is partly funded by the European Union. The DRiVE consortium consists of Ericsson (co-ordinator) BBC, Bertelsmann, Bosch, DaimlerChrysler, Nokia, Steria, Teracom, VCON, and Vodafone as well as Rheinisch-Westfälische Technische Hochschule RWTH Aachen, Universität Bonn, Heinrich-Hertz-Institut Berlin and the University of Surrey. The authors acknowledge the contributions of their colleagues in the DRiVE consortium.

REFERENCES

[Ald01] K.M. ALDINGER, R. KROH, H. FUCHS, *The DRiVE IFA Demonstrator – Test Car and Services*, Mobile Summit 2001, Barcelona, 10–12 September 2001.

[Cam00] A. T. CAMPBELL, J. GOMEZ, *IP Micro-Mobility Protocols*, ACM SIGMOBILE Mobile Computing and Communications Review, Volume 4, Issue 4, October 2000.

[Cas00] C. CASTELLUCCIA, *HMIPv6*, ACM SIGMOBILE Mobile Computing and Communications Review, Volume 4, Issue 1, January 2000.

[Dee98] S. DEERING, R. HINDEN, *Internet Protocol, Version 6 (IPv6) Specification*, IETF RFC 2460, December 1998.

[DRi01] DRiVE project home page: *http://www.ist-drive.org/*

[IFA01] IFA home page: *http://www.ifa-berlin.de*

[Joh01] D. JOHNSON, C. PERKINS, *Mobility Support in IPv6*, Internet Draft <draft-ietf-mobileip-ipv6-14.txt>, IETF mobileip working group, July 2001 (work in progress).

[Lea01] P. LEAVES, S. GHAHERI-NIRI, R. TAFAZOLLI, L. CHRISTODOULIDES, T. SAMMUT, W. STAHL, J. HUSCHKE, *Dynamic Spectrum Allocation in a Multi-radio Environment: Concept and Algorithm*, IEEE Second International Conference on 3G Mobile Communication Technologies, London, United Kingdom, pp. 53–57, 26–28 March 2001.

[Per96] C. PERKINS, *IP Mobility Support*, IETF RFC 2002, October 1996.

[Sol01] H. SOLIMAN, C. CASTELLUCCIA, K. EL-MALKI, *Hierarchical MIPv6 Mobility Management*, Internet Draft <draft-ietf-mobileip-hmipv6-04.txt>, IETF mobileip working group, July 2001 (work in progress).

[Zha98] X. ZHAO, C. CASTELLUCCIA, M. BAKER, *Flexible Network Support for Mobile Hosts*, Proceedings of the Fourth Annual ACM/IEEE International Conference on Mobile Computing and Networking (MobiCom 1998), Dallas, Texas, October 1998.

Chapter 5

From address orientation to host orientation

Pekka Nikander
Ericsson Research Nomadiclab, Jorvas, Finland

Catharina Candolin
Laboratory for Theoretical Computer Science, Finland

Janne Lundberg
Telecommunications Software and Multimedia Laboratory, Finland

1. Introduction

The world is going wireless [GOO 00]. It seems that soon every cell phone, car, and PDA, and even some bicycles, backpacks, jackets, shirts, and shoes will have IP connectivity. We strongly believe that this connectivity will utilize IPv6 [DEE 98], and that the majority of IPv6 hosts will be such small mobile devices. A fundamental difference to the computers of today is that these future devices do not have any natural home location or site where they would spend most of their time. Instead, they tend to create ad hoc networks with physically nearby devices. Often even these ad hoc networks are mobile themselves; for example, a personal network created between your wearable computers usually moves along with you.

In our view, most of the mobile devices will have access to the fixed Internet. However, access is often indirect and intermittent. That is, the ad hoc mobile devices will try to use whatever connections that happen to be available (if any). For example, if one of the devices has a local, fast, and cheap connection (e.g. Wireless LAN), it may act as a router for the nearby devices. On the other hand, if Internet access is only available through a slow and expensive link (e.g. a GSM or a GPRS phone), the connected device may not be that willing to forward packets. In some cases, there might be several simultaneous connections, using different access technologies. We believe that this last case, called multi-access, will become increasingly common.

In this paper, we describe an architecture that provides IPv6 address and mobility management for such future devices and connections. Based on our experimentation, most of what is needed is already there in IPv6. However, by changing IPv6 to be fundamentally host oriented instead of being address oriented, many currently difficult issues get easier. For example, multi homed sites, site renumbering, and mobility management all benefit from our approach. The basic idea in our approach is to add a new slim layer between the IP and the upper layers. This layer provides a mapping between hosts and their current addresses. Each communicating host keeps its active peers aware of any changes in its logical location, thereby allowing connections to survive host mobility and address renumbering. From another point of view, our approach can be described as a generalization of the IETF Mobile IPv6 approach [JOH 00]. In Mobile IPv6, there is a Binding Cache that maps a host identity to its current location (care-of-address). Our approach replaces the Binding Cache with what we call the Host Cache (see Section 3.1). The Host Cache maps a host identity to a set of current locations (a host address set, or multiple care-of-addresses).

The rest of this paper is organized as follows. In the following section we outline the technical background and related work. After that, in Section 3, we describe the new architecture in detail, and especially the differences to the current IPv6 architecture. Section 4 describes the security solutions for our architecture. In Section 5 we discuss the benefits of our architecture, the effect it has on applications, and outline some remaining problems and directions for future work. Finally, Section 6 gives our conclusions from this work, discussing the architectural and other benefits.

2. Background

In the original TCP/IP design and all current implementations, there is a fundamental design decision of binding all connections to IP addresses. However, careful reading of the original TCP specification [POS 80] reveals that TCP connections were meant to be used between hosts, not any particular addresses. The reason for using IP addresses to TCP socket bindings is described in Section 2.7 of [POS 80] as follows: "To provide for unique addresses at each TCP, we concatenate an Internet address identifying the TCP with a port identifier to create a socket which will be unique throughout all networks connected together." Furthermore, the purpose of the TCP pseudo header is to give "the TCP protection against misrouted segments". These two mechanisms also take care of basic security. That is, if we suppose that the network is secure, it is impossible for a TCP implementation to masquerade as another, or a process to masquerade as another process [POS 80].

A similar set of arguments can be found in the UDP specification [POS 81].

Again, the aim is to allow communication between applications, and the applications just happen to be located in hosts, which, in turn, happen to be identified by IP addresses. However, UDP takes a more address oriented approach by defining that a "Destination Port has a meaning within the context of a particular Internet destination address", and by directly using addresses to specify the destination for reply messages.

By analyzing these original requirements, the reasons for using explicit IP addresses in the TCP and UDP specifications can be expressed as follows.

1) IP addresses are used to make TCP and UDP port numbers unique (by combining them with addresses) within the set of all Internet hosts.

2) IP addresses are used to protect TCP and UDP protocol implementations, and the applications above, against misrouted packets.

3) IP addresses are used to prevent malicious TCP implementations from performing masquerade attacks, but only if we can assume that the network itself (the IP layer) is secure.

4) IP addresses are needed to make it possible to answer UDP packets.

5) IP addresses are needed to allow TCP connections to be initiated and to allow initiating UDP packets to be sent.

Of these, only the requirement of being able to reply to UDP packets (point 4) requires that the IP addresses remain fixed for the duration of the transaction or connection. In all other cases, the actual requirement states that the IP address must currently belong to a single predefined host. As we show in Section 3, these requirements can be easily achieved by representing hosts as dynamically changing, non-overlapping sets of IP address; thus, there is no need to assume that each host has one or more fixed, universally assigned addresses.

In this paper, based on the above mentioned observations, we define an architecture where hosts do not necessarily have any fixed IP addresses. That is, there probably will be some hosts with fixed addresses, such as the root DNS servers and possibly some large web servers, but the large majority of hosts will change their addresses more or less dynamically. Many fixed hosts will only go through occasional address changes, e.g., due to router renumbering [CRA 00] (caused by changes in the routing infrastructure) or possibly for privacy reasons [NAR 00]. To them, mechanisms like Dynamic DNS [WEL 00] combined with IPv6 address deprecation might be adequate. However, the truly mobile hosts will change their addresses for mobility reasons, and quite frequently. In our opinion, the truly mobile hosts will be the majority of hosts, and that must be taken care of by the IPv6 architecture.

Thus, since almost all hosts will have more or less dynamic IP addresses, and since there are no real reasons for providing the upper layer protocols with (imaginary) fixed IP addresses, we abandon the concept of fixed IP addresses altogether. Among other things, this makes the home address concept in Mobile

IPv6 obsolete. On the other hand, as mentioned above, it does not mean that all hosts would necessarily have dynamic addresses; it is just that the suggested modification to the IP layer does not make any fundamental distinction between dynamically assigned IP addresses and fixed IP addresses.

Another related issue, not directly covered by the currently presented architecture, is the need for a method of finding out the (current) IP address(es) of a (mobile) host, in particular, of a host that we do not now have anything to do with. In our scheme, the hosts keep track of the addresses of those other hosts that they have active communication with (see Section 3.1). Therefore, it is sufficient that there is some method for querying the addresses when communication between a pair of hosts is first initiated, or if the communication connections were lost or inactive for an extended period of time. For such purposes, Dynamic DNS [WEL 00] or even something like the Session Initiation Protocol (SIP) [HAN 99] seem to be appropriate methods. Thus, in our opinion, there is absolutely no need, not even for connection initiation, for retaining mobile IP home addresses.

2.1 A look at mobility management

Apart from the address vs. host point of view presented above, it is useful to briefly look at the IP mobility problem in general. Mobility in the IP world can be provided at the network-layer or the transport-layer. The basic mobility implementation types seem to fall into the following two categories:

1) **Address translation**. An address translation scheme typically deploys an agent system in the mobile node's home network. The agent intercepts the packets that are destined for the mobile node and forwards the packets to the network where the mobile node is currently located. The forwarding function can be performed by various means, e.g., tunneling. Address translation schemes typically function at the network-layer.

2) **Connection forwarding**. Connection forwarding is typically handled in the transport-layer. When a mobile node changes its IP address, it somehow informs the correspondent host about its new address. The new address can be sent e.g. using TCP-options.

In [BHA 96], Bhagwat, Perkins, and Tripathi discuss mobility at the network-layer. They argue that handling mobility should be completely transparent to protocols and applications running on stationary hosts, and a mobile host should appear like any stationary host connected to the Internet. Many arguments in favor of complete transparency are given in the article. The fundamental reason behind these arguments appears to be unwillingness to alter functionality in existing network infrastructure. Another important reason for transparency is that many existing applications and even protocols above the transport layer assume that a host's IP address never changes during operation. All these applications and protocols would need to be modified. In our opinion, some of their arguments do not hold for IPv6 (see Section 5.2).

On the other hand, Snoeren and Balakrishnan [SNO 00] argue that significant advantage can be gained by handling mobility at the transport-layer because network-layer mobility comes at significant cost, complexity, and performance degradation (due to triangle routing). The motivation for their end-to-end architecture is to support the different mobility modes of applications, that is, applications that originate the connections, and applications to which other hosts originate connections to, and to empower the applications to make choices best suited to their needs.

Our solution is different since it is a connection forwarding scheme working at the network layer, or at the border between the network layer and the transport-layer. As we show later in this paper, our approach has a number of architectural and performance benefits. To our knowledge, there have been few successful attempts to move in the direction we are going. With the exception of the actual Mobile IP approach, most related work presents transport layer solutions (e.g. [SNO 00] [HUI 95] [STE 00b]). One reason for this might be that the original Mobile IP(v4) approach took the overly conservative point of view of assuring full backward compatibility even for existing implementations. In our opinion, full backward compatibility is clearly important for the current IPv4 Internet, but not appropriate to the future IPv6. Based on the discussion above, three possible generic approaches can be identified. All the rest seem to be variations of these. The approaches are enumerated in the following.

1) The current Mobile IPv4 [PER 96] and IPv6 [JOH 00] solution, i.e., a network layer address translation solution. In it an attempt is made to provide full backward compatibility with all upper layer protocols, and to make sure that the upper layer protocols do not need any modifications whatsoever. The selected method is to assume that each and every host has a home address, and that the home address is used to define the connection endpoints. In this solution, the IP headers include both the actual IP addresses, used for routing purposes, plus the home address(es), used for connection endpoint identification purposes. Thus, while providing full backward compatibility, this method results in inefficiencies in the packet size and the utilization of routing infrastructure (triangular routing).

2) In the transport layer oriented connection forwarding solution [SNO 00] [HUI 95] [STE 00b] [WU 97] [STE 00a], the transport protocol is provided with a mechanism that allows a transport endpoint to be associated with a dynamically changing set of IP addresses. While this is an example of sound and efficient engineering, it has the problem of being transport protocol specific. That is, fixing one transport protocol does not help the others.

3) The third possibility, described in this paper, moves the IP layer away from address orientation and towards host orientation. That is, a new layer of data structures is added between the IP layer and the upper layers. The new layer represents each host with a dynamically changing set of IP addresses. Thus, in principle, the solution is pretty similar to the transport layer connection

forwarding case. However, there are two additional benefits from placing the functionality at the IP layer. First, it makes the mechanisms available to all upper layer protocols. Second, it allows the address sets to be shared between multiple transport protocols and multiple connections, thereby saving memory and potentially reducing need for signaling traffic.

2.2 Related work

The transport-layer solution presented by Snoeren and Balakrishnan [BHA 96] share some of our objectives and is discussed in Section 2.2.1. We also briefly describe the SCTP [STE 00a] protocol in Section 2.2.3.

2.2.1 End-to-end mobility

In [SNO 00], a transport-layer mobility scheme is presented. There are three major components in this scheme: addressing, mobile host location, and connection migration. Addresses are obtained using e.g. the Dynamic Host Configuration Protocol (DHCP) [DRO 97], or an autoconfiguration protocol [THO 98].

Once the mobile host has obtained an address, it may start communicating with corresponding hosts. If the mobile host is a client that actively opens a connection to the corresponding host, no location task needs to be performed. However, if the movement occurs while a connection is open, the mobile host will first obtain a new IP address from its new network, and the connection continues via a secure negotiation with the communicating hosts using a Migrate TCP option. To support mobile servers and other applications where hosts residing on the Internet originate the communication, DNS is used for locating mobile hosts. When the mobile host changes its point of attachment, it changes its hostname-to-address mapping in the DNS by using a secure DNS update protocol [EAS 97] [VIX 97]. Hosts wishing to communicate with the mobile host make hostname lookups and retrieve the IP address of the mobile host's current location.

Connection migration is handled by deploying a Migrate TCP option, included in SYN segments, that identifies a SYN packet as part of a previously established connection rather than a request for a new one. The mobile host restarts previously established TCP connections from its new address by sending the Migrate SYN packet that identifies the previous connection; the corresponding host then resynchronizes the connection with the mobile host. To secure the migration from hijacking attacks, the end hosts may either rely on solutions such as IPSec [KEN 98], or use an unguessable connection token which is negotiated using a secret connection key.

The presented scheme is claimed to be secure, efficient, and provide better support for applications to adapt to the host's mobility. However, the scheme is mainly tailored for TCP, and each transport layer protocol would need to be modified to support mobility. Although TCP and UDP are the only widely used transport protocols today, it is likely that there will be several transport protocols in the future. An approach where mobility would have to be incorporated into each available transport protocol is therefore unfeasible, especially since it can be done once and for all on one layer and for one protocol.

Furthermore, the scheme suffers from the so called *double jump* problem, where the two communicating hosts move at the same time. It seems as if no easy solution to this problem exists on the transport-layer, although the authors claim that the problem does not pose a serious limitation to their scheme, since they are primarily targeting infrastructure-based rather than ad hoc network topologies. However, as future mobile devices are likely to form ad hoc networks that have seamless connection to the Internet, and the number of such ad hoc networks is likely to be very large, it seems unfeasible to design transport-layer solutions separately for ad hoc networks and fixed networks as well as transport-layer solutions that work for both ad hoc networks and fixed networks. Therefore, a mobility scheme that does not set any limitations on simultaneous mobility of the hosts and that works in the same fashion on both ad hoc networks and fixed networks is needed.

2.2.2 The LINA architecture

In [ISH 01], the Location Independent Network Architecture (LINA) is presented as an alternative approach to Internet mobility. In IPv6, the IP address typically functions as both an identifier and an interface locator, that is, the address specifies both the identity of the node and its current point of attachment to the Internet. When the node moves, the IP address changes, and its identity is no longer preserved. The main concept of LINA is separation from the node identifier and the interface locator. The node identifier is assigned to the network interface of the node and does not change even if the point of attachment changes. The locator uniquely defines the point of attachment. It is assigned to the network interface of the node and is used for routing purposes.

When an application wishes to communicate with a particular node, it may specify the node using either the identifier or the locator. The transport layer is responsible for maintaining the connection using either one. In the former case, the node identifier must be mapped to its current interface locator. This is done by dividing the network layer into two sublayers: the identification sublayer and the delivery sublayer. The identification sublayer handles the conversion of the identifier to the locator by querying a Mapping Agent (MA), and the delivery sublayer delivers the packet. Mobility support comes from the fact that the node identifier remains unchanged regardless of node movement. In the latter case, where the application wishes to communicate using the locator, the identification sublayer is bypassed, and the delivery layer takes care of routing the packet. However, if either of the communicating nodes move, the transport connection will break. Thus, mobility is not supported in this case.

In [KUN], the LIN6 protocol, which is an application of LINA to IPv6, is described. The main advantages of LIN6 when compared to Mobile IPv6 is smaller header overhead, higher fault tolerance, and the possibility to do end-to-end communication. In this paper, we propose another alternative that is based on Mobile IP but with similar advantages as that of the LINA architecture.

2.2.3 The Stream Control Transmission Protocol

The Stream Control Transport Protocol (SCTP) [STE 00a] is a new IETF standard providing enhanced TCP like transport service. Even though it is primarily designed to provide transport to signalling protocols, e.g., SS7, some people expect that it may replace TCP in some application areas. As a primary feature, the SCTP protocol allows the connection endpoints to be defined as groups of IP addresses. The basic motivation behind this has been the need to provide robustness and load balancing capabilities to large multi-homed hosts [RYT 00]. To further enhance robustness, in their recent Internet Draft [STE 00b], R. R. Stewart et al. propose a method for dynamically adding and removing addresses from the SCTP endpoint address sets. According to Rytina [RYT 00], the actual intention has not been to provide mobility support, but to allow new network interfaces to be added and old ones to be removed from multi-homed hosts.

3. The new architecture

In this section, we describe our modifications and additions to the IPv6 architecture. Technically, the changes are fairly small. However, the changes have fundamental effects on the architecture, since they slightly change the semantics of the API between the IP layer and the upper layers.

First, in Section 3.1, we describe the new basic data structure, the Host Cache. After that, in Section 3.2, we describe how the Host Cache is used to bind the communication endpoints, i.e., the sockets, into hosts rather than single IP addresses, and the effects of this on the transport layer protocols. Section 3.3 briefly describes a new type of socket, Host Sockets, that may be utilized by applications that want to keep track of a host's IP addresses even when there is no active traffic with the hosts. An example of the functionality of our architecture is given in Section 3.4, and our implementation prototype is briefly discussed in Section 3.5.

3.1 Host Cache

The Host Cache is a new fundamental data structure introduced in our architecture. Each host maintains a Host Cache. The Host Cache consists of Host Cache Entries, each describing information about a peer host. An extract from a typical Host Cache is depicted in Figure 1, which contains one local Host Cache entry and two foreign Host Cache entries. The local Host Cache entry contains the information related to the mobile host, such as the IP addresses assigned to it, whereas the foreign Host Cache entries provides information about corresponding hosts.

The most important piece of information stored in a host cache entry is the set

```
Local host
    0000:0000:0000:0000:0000:0000:0000:0001
    fe80:0000:0000:0000:0000:0000:0000:0001
    fe80:0000:0000:0000:0201:02ff:fe33:0fb9
    3ffe:0200:0010:0005:0201:02ff:fe33:0fb9
    fe80:0000:0000:0000:0200:c0ff:fe99:0eac
    3ffe:0200:0010:0005:0200:c0ff:fe99:0eac
```

```
homeless1.nomadiclab.com
    3f80:0200:0000:093c:0200:c0ff:ff99:1cae
    3f80:0200:0000:0952:0200:c0ff:ff30:1f47
```

```
homeless2.nomadiclab.com
    3f80:0200:0000:093c:0200:c0ff:ff99:1cae
    3f80:0200:0000:0952:0200:c0ff:ff30:1f47
```

Figure 1. *Typical Host Cache contents*

of IP addresses through which the host is currently reachable. For a simple, fixed, non-multihomed host the cache entry simply contains a single address, the address associated with the host. Correspondingly, an entry describing a fixed multi homed host contains a fixed set of addresses, and the entry describing a mobile host contains a dynamic set of addresses. Each address is associated with a routing entry, giving information about the reachability of the address, address lifetime, giving an expiration time for the address, and possibly other information such as which address the host prefers to use right now.

A cache entry is created, for example, when a client initiates a connection to the host, or as a result of a reply received to a DNS query. Upon the creation of the cache entry, it is seeded with information about the assumed addresses of the host.

Once a cache entry is created, it is the responsibility of the peer host, i.e., the host represented by the entry, to keep the entry up to date. The peer takes care of this by sending Mobile IPv6 Binding Updates. Thus, from a Mobile IPv6 point of view, a Binding Update does not any more update the single binding between a home address and a care-of-address, but it adds or deletes addresses in the host address set. This allows the corresponding hosts to keep up to date information about all the addresses of a mobile host instead of just using a single care-of-address. Furthermore, no explicit home addresses are needed. To identify a cache entry, any existing address within a cache entry can be used to identify the particular entry.

As almost all IPv6 addresses have some kind of lifetime, e.g., as a result of acquiring them through the IPv6 duplicate address detection process [THO 98], it is natural that all addresses in the Host Cache also have a lifetime. Once the lifetime of an address expires, the address is marked deprecated, and it is no longer used when sending packets. However, even deprecated addresses may be used when handling incoming packets.

Each host is individually responsible for keeping its active peers aware of the lifetimes of its addresses. For that reason, Binding Updates may be sent not only to add or remove addresses in the peer's host cache, but also to inform the peer about changes in the lifetimes of active addresses. For security reasons, all updates to the host caches must be authenticated. Security is discussed further in Section 4.

3.2 Socket bindings

Basically, there is just one architectural change in our solution: we add a slim data structure layer between the transport layer protocols and the IP layer. In practice, this change is most clearly visible in the way sockets, i.e., communication endpoints, are bound in the kernel data structures.

Figure 2 depicts the standard way of binding sockets. In the current standard solution, a connected socket is bound to a single local and a single foreign address. Whenever sending data through the socket, the local address is used as the source address for packets, and the foreign address is used as the destination address. Respectively, when receiving packets, only such packets are routed to the socket where the source address is equal to the foreign address and the destination address is equal to the local address.

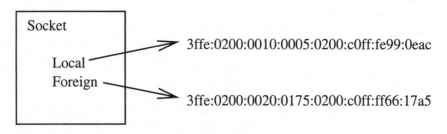

Figure 2. Socket bindings in standard IPv6 implementation

Figure 3 depicts our modified data structures. Here, a socket is not bound to single addresses but Host Cache Entries. A local Host Cache Entry enumerates the local addresses which the socket may use. Correspondingly, a foreign Host Cache Entry enumerates the addresses of the foreign host that we know and may use to reach the host.

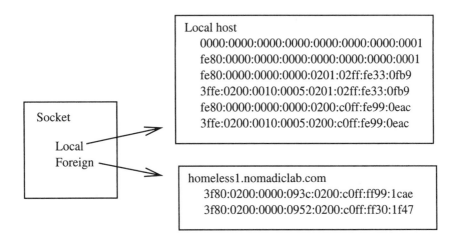

Figure 3. Socket bindings in our modified IPv6 architecture

Usually the foreign Host Cache Entries are shared between all sockets that are bound to the same host. Correspondingly, there is usually just one local Host Cache Entry, containing all the local IP addresses. However, there is no reason why there could not be more local Host Cache Entries than just one. This might be useful, for example, in large servers where the number of concurrently open sockets exceeds the 16 bit integer range, or to provide a different set of addresses for trusted connections and another set of addresses for untrusted connections. If there is more than one local Host Cache Entry, care must be taken that they stay mutually exclusive with respect to the set of foreign hosts they serve.

When a packet is sent, the sending host must select one of the addresses to be used as a destination address. The choice is made based on some specified policy. For example, the policy may state that the destination address to be used shall be the last address that the peer host used as a source address. Another possible policy would be to use the *fattest pipe*, that is, the destination address corresponding to the link with the highest speed.

The selection may also be made on a packet basis. This may be beneficial; for example, this can be used for load balancing between links, or selecting the link based on some Quality-of-Service policy or other properties. Once the destination address is selected, the source address may be selected among the ones available at the local Host Cache Entry, for example, using the source address selection algorithm in [DRA 01].

Similarly, when a packet is received its source and destination addresses are matched against the Host Cache. If there is a foreign Host Cache Entry that contains the source address, and a corresponding local Host Cache Entry that contains the destination address, it is easy to determine if there is a socket bound to the entries found and the ports present in the packet. Similarly, if there is no such bound socket, it is simple to look for a listening socket that would be willing to take care of the packet.

In practice, though, the situation is slightly more complex. Some care must be taken, since under some configurations it is possible that the destination address is found in several local Host Cache Entries. If a bound socket is found, there are no real problems since there may only be one socket that is bound to the foreign Host Cache Entry and the ports. Thus, the socket bindings uniquely determine which of the matching local Host Cache Entries is the right one. However, if there is no such bound socket, the selection among the matching local Host Cache Entries is not straightforward anymore. Basically, some local policy rule must be applied, using the foreign address and the foreign Host Cache Entry (if any) as input. In the default case, the policy would state that if there are any sockets bound to the foreign Host Cache Entry, take the same local Host Cache entry that is shared among those sockets. However, there are reasons for more complex policies. For example, to achieve real load balancing, to handle anycast addresses, or due to security policy reasons, the policy may be more complicated.

3.3 Host sockets

There are situations where a host wants to actively keep track of the addresses of a foreign host even though there is no ongoing communication between them. For example, if the host has received a Jini lease [ARN 99] from a foreign host, it needs to keep track of the foreign host in order to be able to renew the lease at a later time. For this purpose, we introduce a new concept of host sockets. A host socket is a socket in the sense that it is bound to a local and foreign host cache entry as the regular bound sockets are. However, a host socket does not allow any data to be sent between the hosts. Its only purpose is to keep the foreign host cache entry active, thereby enforcing the IP layer to send Binding Updates whenever there are changes among the set of local IP addresses.

3.4 Example

The functionality of our approach is best described with an example. The example is divided into host originated and host terminated connections, that is, in the former the host contacts another host on the Internet, whereas in the latter the host is contacted by a remote host. Furthermore, a solution to the double jump problem is proposed.

3.4.1 Host originated connections

When a host wishes to establish a connection e.g. to a server, it requests the IP address(es) of the server from DNS and creates a foreign Host Cache entry based on the response. After that, it chooses, according to the specified policy, the source and destination address to which the socket is to be bound, and establishes the connection to the server.

If the host roams during the connection, it must inform the corresponding host about the change of point of attachment. This is done by updating its Host Cache entry at the corresponding host.

3.4.2 Host terminated connections

Mobile hosts functioning as servers need a way to announce their current location to the outside world. One possible way would be to make use of dynamic DNS [WEL 00]. When the host changes its point of attachment, it updates its name-to-address entries in the DNS in a similar fashion as in the transport-layer solution presented by [SNO 00] (discussed in Section 2.2.1). A host originating a connection will thus find the host by querying the DNS, which returns one or more addresses through which the host can be currently reached. The connection originating host then creates a Host Cache entry corresponding to the DNS replies and uses one of the addresses according to its policy.

The major drawback of using Dynamic DNS is, however, that it does not support fast handoffs where the host changes its address. However, if we assume that a host does not change its point of attachment to the Internet more frequently than a few times a minute, the scheme should work well. This assumption is made by several other mobility schemes as well and seems to be feasible in most cases. However, our solution is by no means restricted to Dynamic DNS; other mechanisms, such as SIP [HAN 99], may be deployed instead.

3.4.3 Addressing the double jump problem

One of the problems our architecture does not currently address is the so called double jump problem [HUI 95]. Basically, the double jump problem appears when two mobile hosts move at the same time. As a consequence, both of them send Binding Updates, but to the old addresses. Thus, clearly, a mechanism is needed to forward the packets from their old locations to the new ones, or the connections will break.

Fortunately, the standard Mobile IPv6 [JOH 00] provides an easily deployable solution. In Section 10.9 of [JOH 00], a mechanism is specified where a mobile host arranges a home agent at the old location to temporarily forward packets to its new location. Our intention is to use this mechanism. However, from an architectural point of view, we would prefer to rename that kind of agents to forwarding agents, instead of calling them home agents. That is, in our architecture, the purpose of those agents is to temporarily forward packets, not to act as some kind of home for the mobile hosts. When the host moves to another network, it will register its new address to the forwarding agent, which for a short period of time (e.g. 15 seconds) will forward the packets to the host. The forwarding request must be authenticated in order to prevent illegitimate requests. Authentication is discussed in Section 4.

3.5 Implementation

We have implemented a prototype based on FreeBSD 4.2-STABLE with the KAME IPv6 stack. In this section, we will briefly mention some of the most fundamental modifications made to the BSD TCP/IP stack as well as to the KAME IPv6 implementation.

The Host Cache is based on an almost complete rewrite of the existing FreeBSD host cache implementation. Our implementation consists of a generic host cache module and an IPv6 specific module. The latter utilizes the services provided by the former. The generic module is based on two major data structures; the host cache entries themselves and host cache addresses. A host cache entry contains a number of host cache addresses, together with some other data. The IPv6 host cache module adds additional information and services to the host cache entries.

We also made some modifications to the Internet protocol control block in the BSD/KAME stack, augmented the TCP code with dynamic source and destination address selection, and made some small changes to the optimized TCP route handling mechanism. We also modified the KAME API to take or return socket addresses instead of plain IP addresses. Some other minor changes were made to the BSD and the KAME stack, but the details are out of the scope of this paper.

3.5.1 Deployment

At the present stage, the protocol has not been deployed, and no deployment is planned for the presented version. Instead, we are looking at how some of the presented ideas could be combined with the Host Identify Payload (HIP) [MOS 01].

4. Security

As discussed in the previous section, a mechanism for authenticating updates to the Host Cache as well as forwarding requests is needed. Our solution is based on public key cryptography; thus, each host has in its possession a public/private cryptographic key pair that is used to identify the host. To support the minimal desired level of protection in our architecture, the following security goals must be met:

1) IP addresses are unsuitable for security purposes and should therefore only be used to provide reachability information.

2) There must be at least some assurance that, when we encounter a new host, we are actually able to communicate with the host using the given IP address and that the host actually has the private key associated to the given public key in its possession.

3) Third, whenever the host changes its set of IP addresses, we want to have at least as much assurance about the authenticity of the change as we have about the initial binding between the key and the addresses.

When two hosts that have no prior knowledge of each other wish to establish a connection, they will exchange public keys and prove that they are in possession of the corresponding private keys. The public key of a foreign host will be stored together with the Host Cache entry representing that host. However, the key is otherwise unrelated to the set of IP addresses in the cache entry. The IP addresses merely provide reachability information for the host that is identified by the key that was given when the cache entry was initially created. The IP addresses are thus free to change over time, as long as the changes are made by the host that is in possession of the corresponding private key.

When the foreign host roams to another network, and thus needs to update its cache entry, or e.g. when a host socket is used, a Binding Update will be sent to the local host. The local host verifies that the Binding Updates are legitimate, and makes the requested updates to the corresponding host cache entry.

Authentication of forwarding requests is based on the same principle. When the host first enters the network of the forwarding agent and obtains an IP address, a key exchange will take place between them. Should the host later roam to another network and request the forwarding agent to send packets destined for it to its new location, it will prove that it is in possession of the private key corresponding to the public key that it presented when it obtained its IP address at the network of the forwarding agent. After that, the forwarding agent will forward packets destined for the host for a short period of time.

5. Discussion

In this section, we discuss the advantages of our scheme (Section 5.1) as well as the implications it has on existing applications (Section 5.2). Finally, some remaining problems and open areas are discussed (Section 5.3).

5.1 Benefits

The main benefit of our architecture is that we have been able to eliminate the need for a home address for hosts. Thus, our scheme avoids the performance degradation problems caused by triangle routing, since the hosts do not have a dedicated home network to which the packets are routed.

Another major benefit of our architecture is the reduced average header size. When standard Mobile IPv6 [JOH 00] is used between two mobile hosts, all packets carry two extension headers: the routing header and the home address destination option. Together, these two headers usually take 52 bytes, increasing the IPv6 header size from 40 bytes to 92 bytes. In our scheme, no extensions are needed. Additionally there are no big differences in the signalling load.

Furthermore, since transport layer connections are no longer bound to interfaces, a host can choose the interface it uses to transmit the packets. Thus, the host can transmit the packet through the interface which offers, for example, the cheapest service, the smallest round-trip-time, or the best throughput.

5.2 Effects on applications

Most applications never detect the proposed change of semantics in the underlying IP layer. They open a connection, i.e., bind a socket, using whatever IP address the local DNS resolver gives them. After that they just forget about the address, and happily use the connection. If the underlying addresses change, the application is completely unaware of that as long as the connection continues to function as before. We call this class of applications address agnostic.

Unfortunately, there are a few classes of applications that are not that address agnostic. First, many UDP servers use unconnected UDP sockets, and compose reply packets by themselves, based on the addresses received in the query. Second, there is a small class of applications that send IP addresses within protocol packets. FTP is the prime example of such an application. And third, there are some applications that store IP addresses in long term storage, or otherwise explicitly handle IP addresses.

According to our understanding, most UDP applications should work normally even if the underlying IP addresses change. Typically, the UDP transactions are relatively short term, and in the rare case that the reply is completely lost, they rely on error recovery mechanisms built on the top of UDP. On the other hand, some UDP applications use IP addresses for filtering. That is, they use the source IP address to make a decision if an operation is permitted or not. In our opinion, IP addresses should not be relied on for security purposes, and therefore this latter class of applications should be redesigned anyway.

The protocols that include IP addresses inside upper layer protocol packets may or may not suffer from the changing addresses. In principle, even FTP could survive since the IP address is needed only temporarily when opening a new transfer connection. However, in practice few if any FTP clients request their current IP address individually for every PORT command, but instead store the address and reuse it whenever needed. Thus, in practice, most of the applications in this category would need work. On the other hand, most of these applications will need to be reworked anyhow in the process of adapting them to work in the IPv6 world.

The third class of applications, i.e., those that store IP addresses for extended periods of time, are in clear conflict with the Internet Architecture (RFC 1958 [CAR 96], Section 4.1). Therefore, they should be redesigned anyway.

Thus, in summary, we believe that all well designed applications will continue unaffected even when running on the top of the new architecture. Coming to the exceptions, it is our strong opinion that their design flaws should be fixed in order to be compliant with the Internet Architecture and sound security principles. Once that is done, even those applications should work well.

On the negative side, one of the functions that would break is packet filtering based on IP addresses. This covers both filtering performed at firewalls and routers as well as filtering mechanisms at the hosts. But, if route optimization in Mobile IP will ever become common, much of the current firewall and router packet filtering mechanisms are likely to break anyway. What comes to the host packet filtering mechanisms, they could be fairly easily fixed by replacing individual IP addresses in the filters with host cache entries.

5.3 Open issues and future work

In Section 5.3.1 we discuss the problem of backward compatibility and describe how we currently solve it partially. Section 5.3.2 discusses the problem of supporting mobile networks. Finally, Section 5.3.3 lists a number of minor issues that need to be fixed before our approach is ready for prime time.

5.3.1 Full backward compatibility

Backward compatibility may be divided into three issues: backward compatibility with standard IPv6 hosts, backward compatibility with hosts implementing the IETF Mobile IPv6 specification, and the problem of providing full backward compatibility for old badly behaving applications. Currently we leave this final problem, i.e., the troublesome application backward compatibility one, for future work, and only discuss the other two questions.

Providing backward compatibility to standard IPv6 hosts is fairly straightforward. Basically, whenever a host realizes that the peer does not accept Binding Updates (see Section 3.1), it marks the Host Cache Entry of the peer appropriately, and reverts back to the old socket semantics whenever communicating with that host. As a consequence, connections will remain connected as long as neither of the hosts move.

Backward compatibility with hosts compliant with the current IETF Mobile IPv6 specification requires some more care. In our current implementation, we add a new suboption, "Homeless support", to the Binding Request and Binding Update options. The presence or absence of this option allows the hosts to trivially determine whether the peer supports the new semantics proposed by us, or the standard IETF Mobile IPv6 semantics.

5.3.2 Mobile networks

A mobile network, as opposed to a mobile host, is a network that changes its connectivity points with the fixed Internet. Examples of expected future mobile networks include personal, airplane, and car networks, to mention but a few. As discussed by Ernst et al. in [ERN 00], the current IETF Mobile IPv6 proposal does not seem to be an optimal solution to support such mobile networks. An interesting future direction is to see if our scheme can be extended to efficiently support mobile networks in addition to mobile hosts.

5.3.3 Some minor issues

Based on our experimentation, there seems to be a number of minor problems that are more specific to mobility in general rather than to our solution. First, there is clearly need for an API where applications can determine how long a TCP session should wait for the peer to become on-line again before giving up. This API should probably be somehow combined with better control over TCP keepalives. Second, there is a clear need for the IP layer to signal the TCP layer whenever the preferred addresses are changed. This would allow the TCP to take a more efficient approach to congestion avoidance. A still better solution would probably move some of the data currently held in TCP protocol control blocks into the host cache entries, even annotating the individual addresses. In the light of the new architecture, the TCP and UDP socket APIs would probably benefit from some changes. For example, it would be beneficial if, e.g., the connect system call would take several addresses instead of just a single one. Even better would probably be a higher level API that would hide the name resolving step from the application, allowing the application to directly operate on names.

6. Conclusions

In this paper, we have proposed a small but fundamental modification to the IPv6 architecture, and described our prototype implementation. The benefits of the modification are twofold. First, it allows a number of difficult problems, including mobility, multi-homing, site renumbering, and simultaneous multi-access, to be solved in a unified, architecturally elegant, and secure way [LUN 01] [NIK 01]. Second, it reduces the average Mobile IPv6 packet header size.

Architecturally, our method allows IP addresses to be used purely for reachability purposes. Addresses are still used for connection identification, too. As a fundamental difference, though, the connections are bound to dynamically changing sets of addresses instead of fixed single addresses. This approach has the additional benefit that the wire protocols may be preserved unmodified.

From an implementation point of view, our approach adds a thin host representation layer between the upper layers and the network layer. This layer represents the hosts as address sets, and allows changes to the address sets to be communicated between hosts. From the performance viewpoint, the basic advantage of our method, compared to the standard IETF Mobile IPv6 [JOH 00], is the decrease in average IPv6 header size.

Thus, as a summary, the proposed approach makes many of the end-to-end problems in the current Internet easier to solve, is sound with the Internet Architecture, better reflects the needs of the future Internet, and requires a significantly smaller average header size than the current Mobile IPv6 standard, and can be incrementally deployed.

Acknowledgements

We present our thanks to the members of the CAP project group, without whom we would not have tackled this problem [NIK 00]. We also want to thank Tuomas

Aura for working as a coauthor on the parallel Internet Draft [NIK 01]. Our thanks also belong to Garett Wollman of MIT, who wrote the original FreeBSD host cache code back in 1997 (though for a completely different purpose than for what we use it), and to all the people that have contributed in a form or another to FreeBSD and KAME.

REFERENCES

[ARN 99] Arnold K., Scheifler R., Waldo J., Wollrath A., O'Sullivan B., *The Jini Specification*, Addison-Wesley, June 1999, ISBN 0201616343.

[BHA 96] Bhagwat P., Perkins C., Tripathi S., "Network Layer Mobility: An Architecture and Survey", *IEEE Personal Communications*, 1996.

[CAR 96] Carpenter B., "Architectural Principles of the Internet", *RFC 1958*, Internet Architecture Board, June 1996.

[CRA 00] Crawford M., "Router Renumbering for IPv6", *RFC 2894*, IETF, August 2000.

[DEE 98] Deering S., Hinden R., "Internet Protocol, Version 6 (IPv6) Specification", *RFC 2460*, IETF, December 1998.

[DRA 01] Draves R., "Default Address Selection for IPv6", Internet draft, draft-ietf-ipngwg-default-addr-select-04.txt, IETF, March 2001.

[DRO 97] Droms R., "Dynamic Host Configuration Protocol", *RFC 2131*, IETF, March 1997.

[EAS 97] Eastlake 3rd D., "Secure Domain Name System Dynamic Update", *RFC 2137*, IETF, April 1997.

[ERN 00] Ernst T., Bellier L., Claude C., H.-Y. L., "Mobile Networks Support in Mobile IPv6", Internet Draft, draft-ernst-mobileip-v6-network-00.txt, work in progress, July 2000.

[GOO 00] Goodman D., "The Wireless Internet: Promises and Challenges", *IEEE Computer*, vol. 33, num. 7, 2000.

[HAN 99] Handley M., Schulzrinne H., Schooler E., Rosenberg J., "SIP: Session Initiation Protocol", *RFC 2543*, IETF, March 1999.

[HUI 95] Huitema C., "Multi-homed TCP", Internet Draft (expired), IETF, May 1995.

[ISH 01] Ishiyama M., Kunishi M., Uehara K., Esaki H., Teraoka F., "LINA: A New Approach to Mobility Support in Wide Area Networks", *IEICE Transactions on Communications*, 2001.

[JOH 00] Johnson D., Perkins C., "Mobility Support in IPv6", work in progress, Internet Draft, draft-ietf-mobileip-ipv6-12.txt, IETF, April 2000.

[KEN 98] Kent S., Atkinson R., "Security Architecture for the Internet Protocol", *RFC 2401*, IETF, November 1998.

[KUN] Kunishi M., Ishiyama M., Uehara K., Esaki H., Teraoka F., "LIN6: A New Approach to Mobility Support in IPv6".

[LUN 01] Lundberg J., "Infrastructureless IP Mobility for Multi-homed Hosts", Master's thesis, Helsinki University of Technology, February 2001.

[MOS 01] Moskowitz R., "Host Identity Payload Architecture", Internet Draft, draft-ietf-moskowitz-hip-arch-02.txt, work in progress, 2001.

[NAR 00] Narten T., Draves R., "Privacy Extensions for Stateless Address Autoconfiguration in IPv6", Internet Draft, draft-ietf-ipngwg-addrconf-privacy-03.txt, work in progress, IETF, September 2000.

[NIK 00] Nikander P., "Combining Trust Management, Jini, IPv6 and Wireless Links (Extended Abstract)", Tim Kindberg F. M., Posegga J., Eds., *Infrastructure for Smart Devices – How to Make Ubiquity an Actuality*, Bristol, September 2000, HUC.

[NIK 01] Nikander P., Lundberg J., Candolin C., Aura T., "Homeless Mobile IPv6", work in progress, Internet Draft, draft-nikander-mobileip-homelessv6-01.txt, IETF, February 2001.

[PER 96] Perkins C., "IP Mobility Support", *RFC 2002*, IETF, October 1996.

[POS 80] Postel J., "Transmission Control Protocol", *RFC 793*, IETF, January 1980.

[POS 81] Postel J., "User Datagram Protocol", *RFC 768*, IETF, September 1981.

[RYT 00] Rytina I., November 2000, private communications.

[SNO 00] Snoeren A., Balakrishnan H., "An End-to-End Approach to Host Mobility", *Proceedings of ACM MobiCOM 2000*, ACM, 2000.

[STE 00a] Stewart R., Xie Q., Morneault K., Sharp C., Schwarzbauer H., Taylor T., Rytina I., Kalla M., Zhang L., Paxson V., "Stream Control Transmission Protocol", *RFC 2960*, IETF, October 2000.

[STE 00b] Stewart R., Xie Q., Tuexen M., Rytina I., "SCTP Dynamic Addition of IP Addresses", Internet Draft, draft-stewart-addip-sctp-sigtran-01.txt, work in progress, IETF, November 2000.

[THO 98] Thomson S., Narten T., "IPv6 Stateless Address Autoconfiguration", *RFC 2462*, IETF, December 1998.

[VIX 97] Vixie P., Thomson S., Rekhter Y., Bound J., "Dynamic Updates in the Domain Name System (DNS UPDATE)", *RFC 2136*, IETF, April 1997.

[WEL 00] Wellington B., "Secure Domain Name System (DNS) Dynamic Update", *RFC 3007*, IETF, November 2000.

[WU 97] Wu I., Zhang B., "Extended Transmission Control Protocol (ETCP) Project", University of California, Berkeley, http://www-ieee.eecs.berkeley.edu/~irenewu/ETCP/, December 1997.

Chapter 6

Performance analysis of OLSR multipoint relay flooding in two ad hoc wireless network models

Philippe Jacquet, Anis Laouiti, Pascale Minet and Laurent Viennot

INRIA, Rocquencourt, Le Chesnay, France

1. Introduction

Radio networking is emerging as one of the most promising challenges made possibly by new technology. Mobile wireless networking brings a new dimension of freedom in Internet connectivity. Among the numerous architectures that can be adapted to radio networks, the *ad hoc* topology is the most attractive since it consists of connecting mobile nodes without pre-existing infrastructure. When some nodes are not directly in range of each other there is a need for packet relaying by intermediate nodes. The working group MANet of Internet Engineering Task Force (IETF) is standardizing routing protocol for *ad hoc* wireless networking under Internet Protocol (IP). In MANet every node is a potential router for other nodes. The task of specifying a routing protocol for a mobile wireless network is not a trivial one. The main problem encountered in mobile networking is the limited bandwidth and the high rate of topological changes and link failure caused by node movement. In this case the classical routing protocols such as Routing Internet Protocol (RIP) and Open Shortest Path First (OSPF) first introduced in ARPANET [1] are not adapted since they need too much control traffic and can only accept few topology changes per minute.

The MANet working group proposes two kinds of routing protocols:

1. the reactive protocols;
2. the pro-active protocols.

The reactive protocols such as AODV [3], DSR [2], and TORA [4], do not need control exchange data in absence of data traffic. Route discovery procedure is invoked on demand when a source has a new connection towards a new destination pending. The route discovery procedure in general consists of the flooding of a *query* packet and the return of the route by the destination. Exhaustive flooding can be very expensive, thus creating delays in route establishment. Furthermore, the route discovery via flooding does not guarantee creation of optimal routes in terms of hop-distance.

The pro-active class gathers protocols that need periodic updates with a control packet and therefore generate extra traffic which adds to the actual data traffic. Examples of such protocols are Fish EYES Routing (FSR [5]), Optimized Link State Routing (OLSR [6]), Topology Based Reverse Path First (TBRPF [7]). In these protocols the control traffic is broadcast all over the network via optimized flooding. Optimized flooding is possible since nodes permanently monitor the topology of the network. OLSR uses multipoint relay flooding which very significantly reduces the cost of such broadcasts. Furthermore, the nodes have a permanent dynamic database which makes optimal routes immediately available on demand. The protocol OLSR has been adapted from the *intra-forwarding* protocol in HIPERLAN type 1 standard [8]. Most of the salient features of OLSR, such as multipoint relays and link state routing, already exist in the HIPERLAN standard.

The aim of the present paper is to analyze the performance of the multipoint relaying concept of OLSR using two models of network: the random graph model and the random unit graph model. The paper is divided into four main sections. The first section summarizes the main feature of OLSR protocol. The second section introduces and discusses the graph models. The third section develops the performance analysis of OLSR with respect to the graph models. An internal research report exists [15].

2. The Optimized Link State Routing protocol

2.1 Non optimized link state algorithm

Before introducing the optimized link state routing we briefly recall the non optimized link state such as OSPF. In an *ad hoc* network, a pair of two nodes which can hear each other are called a *link*. In order to achieve unicast transmission, it is important here to use a bidirectional link (IEEE 802.11 radio LAN standard requires a two way packet transmission). However due to sensitivity of power discrepancies, unidirectional links can arise in the network. The use of unidirectional links is possible but requires different protocols and is omitted here. Each link in the graph is a potential hop for routing packets. The aim of a link state protocol is that each node has sufficient knowledge about the existing link in the network in order to compute the shortest path to any remote node.

Each node operating in a link state protocol performs the following two tasks:

- **Neighbor discovery**: to detect the adjacent links;
- **Topology broadcast**: to advertize over the whole network important adjacent links.

By important adjacent links we mean a subset of adjacent links that permit the computation of the shortest path to any destination.

The simplest neighbor discovery is for each node to periodically broadcast full hello packets. Each full hello packet contains the list of the heard neighbor by the node. The transmission of hello packets is limited to one hop. By comparing the list of heard nodes each host determines the set of adjacent bidirectional links.

A non optimized link state algorithm performs topology broadcast simply by periodically flooding the whole network with a topology control packet containing the list of all its neighbor nodes (i.e. the heads of its adjacent links). In other words, all adjacent links of a node are important. By flooding we mean that every node in the network re-broadcasts the topology control packet upon reception. Using a sequence number prevents the topology control packet being retransmitted several times by the same node. The number of transmission of a topology control packet is exactly N, when N is the total number of nodes in the network, when retransmission and packet reception are error free.

If h is the rate of hello transmission per node and τ the rate of topology control generation, then the actual control overhead in terms of packet transmitted of OSPF is

$$hN + \tau N^2 \tag{1}$$

As the hello packets and the topology control packet contain the IP addresses of originator node and neighbor nodes, the actual control overhead can be expressed in terms of IP addresses unit. The overhead in IP address units is

$$hNM + \tau N^2 M \tag{2}$$

where M is the average number of adjacent links per node. If M is of the same order as N then the overhead is cubic in N. Notice that the topology broadcast overhead is one order of magnitude larger than the neighbor discovery overhead.

Notice that for non-optimized link state routing the hello and topology control packet can be the same.

2.2 OLSR and MultiPoint Relay nodes

The Optimized Link State Routing protocol is a link state protocol which optimizes the control overhead via two means:

1. the important adjacent links are limited to MPR nodes;
2. the flooding of the topology control packet is limited to MPR nodes (MPR flooding).

The concept of MultiPoint Relay (MPR) nodes has been introduced in [8]. By MPR set we mean a subset of the neighbor nodes of a host which covers the two-hop neighborhood of the host. The smaller the MPR set the more efficient will be the optimization. We give a more precise definition of the multipoint relay set of a given node A in the graph. We define the neighborhood of A as the set of nodes which has an adjacent link to A. We define the two-hop neighborhood of A as the set of nodes which has an invalid link to A but has a valid link to the neighborhood of A. This information about two-hop neighborhoods and two-hop links is made available in hello packets, since every neighbor of A periodically broadcasts their adjacent links. The multipoint relay set of A (MPR(A)) is a subset of the neighborhood of A which satisfies the following condition: every node in the two-hop neighborhood of A must have a valid link towards MPR (A).

The smaller the Multipoint Relay set is, the more optimal is the routing protocol. [13] gives an analysis and examples about multipoint relay search algorithms. The MPR flooding can be used for any kind of long hole broadcast transmission and follows the following rule:

A node retransmits a broadcast packet only if it has received its first copy from a node for which it is a multipoint relay.

[8] provides proof that such flooding protocol (selective flooding) eventually reaches all destinations in the graph. [8] also gives proof that for each destination in the network, the subgraph made of all MPR links in the network and all adjacent links to host A contain a shortest path with respect to the original graph.

Therefore the multipoint relays improve routing performance in two aspects:

1. it significantly reduces the number of retransmissions in a flooding or broadcast procedure;

2. it reduces the size of the control packets since OLSR nodes only broadcast its multipoint relay list instead of its whole neighborhood list in a plain link state routing algorithm.

In other words if D_N is the average number of MPR links per node and R_N the average number of retransmission in an MPR flooding, then the control traffic of OLSR is, in the packet transmitted:

$$hN + \tau R_N N, \tag{3}$$

and, in IP addresses transmitted:

$$hMN + \tau R_N D_N N. \tag{4}$$

Notice that when the nodes selects all their adjacent links as MPR links, we have $D_N = M$ and $R_N = N$: we have the overhead of a full link state algorithm. However we will show that straightforward optimizations make $D_N \ll M$ and $R_N \gg N$ gaining several orders of magnitude in topology broadcast overhead. Notice that the neighbor discovery overhead is unchanged. Summing both overheads we may expect that OLSR has an overhead reduced by a magnitude order with respect to full link state protocol.

The protocol as proposed in IETF may differ in some details from this very simple presentation. The reason is to obtain second order optimization with regards to mobility for example. For example hosts in actual OLSR do not advertize their MPR set but their MPR selector set, i.e. the subset of neighbor nodes which have selected this host as MPR.

2.3 MPR selection

Finding the optimal MPR set is an NP problem as proved in [9]. However there are very different heuristics. Amir Qayyum [13] has proposed the following one:

1. select as MPR the neighbor node which has the largest number of links in the two-hop neighbor set;
2. remove this MPR node from the neighbor set and the neighbor nodes of this MPR node from the two-hop neighbor set;
3. repeat the previous steps until the two-hop neighbor set is empty.

An ultimate refinement is a prior operation which consists of detecting in the two-hop neighbor the node which has a single parent in the neighbor set. These parents are selected as MPR and are eliminated from the neighbor set, and their neighbor are eliminated in the two-hop neighbor set.

It is proven in [9, 11] that this heuristic is optimal by a factor $\log M$ where M is the size of the neighbor set (i.e. the heuristical MPR set is at most $\log M$ times larger than the optimal MPR set).

2.4 Deterministic properties of OLSR protocol

The aim of this section is to show some properties of OLSR protocol which are independent of the graph model. Basically we will prove the correct functioning of OLSR protocol. In particular we will prove that the MPR flooding actually reaches all destinations and that the route computed by OLSR protocol (and actually used by data packets) has optimal length.

To simplify our proof we call chain of nodes any sequence of nodes $A_1,...A_n$ such that each pair (A_i, A_{i+1}) are connected by a (bidirectional) link.

Theorem 1 *When at each hop broadcast packets are received error-free by all neighbor nodes, the flooding via MPR reaches all destinations.*

Note
When the transmissions are prone to errors, there is no guarantee of correct delivery of the broadcast packet to all destinations, even with a full flooding retransmission process. Amir Qayyum, Laurent Viennot Anis Laouiti *et al.* show

in [11] the effect of errors on full flooding and MPR flooding. It basically shows that MPR flooding and full flooding have similar reliability.

Proof: Since we assume error free broadcast transmission, any one hop broadcast reaches all neighbor nodes.

Let us assume a broadcast initiated by a node A. Let B be an arbitrary node in the network. Let k be smallest number of hops between B and the set of nodes which eventually receive the broadcast message. We will show that $k = 0$; i.e., B receives the broadcast message.

Let us assume *a contrario* that $k > 0$. Let F be the first node at distance $k + 1$ from node B which retransmitted the broadcast message. We know that this node exists since there is a node at distance k from node B which received the broadcast message. Let $F_1,..., F_k$ be the chain of nodes that connects node B to node F in $k + 1$ hops. Node F_{k-1} is at distance $k - 1$ from node B (when $k = 1$, node F_{k-1} is node B). Node F_{k-1} is also in the two-hop neighborhood of node F. Let F' be an MPR of node F which is a neighbor of F_{k-1}. Since node F' receives its first copy of the broadcast message from its MPR selector F, F' must retransmit the broadcast message, which contradicts the definition of k.

We now prove that the route computed by OLSR protocol has optimal length. It is easy to see that this property is the corollary of the following theorem.

Theorem 2 *If two nodes A and B are at distance $k + 1$, then there exists a chain of nodes $F_1,..., F_k$ such that the three following points hold:*

1. node F_k is MPR of node B;
2. node F_i is MPR of node F_{i+1};
3. node F_1 is a neighbor node of node A.

Proof: The proof is by induction. The property is trivial when $k = 0$. When $k = 1$, node A is in the two-hop neighborhood of B. Thus there exists an MPR node F of node B which is at distance one hop of node A, which proves the property for $k = 1$. Let us now assume that the property is true up to a given value k. We will prove that the property is also true for value $k + 1$. Let assume a node B at distance $k + 2$ of node A. There exists an MPR node F of node B which is at distance $k + 1$ of A (for verification, there exists a node F' which is at distance k from A and at distance 2 from B and node F can be one of the MPR nodes of B that cover F'). By the recursion hypothesis there exists a chain $F_1,..., F_k$ which connects A to F such that:

1. node F_k is MPR of node F;
2. node F_i is MPR of node F_{i+1};
3. node F_1 is a neighbor node of node A.

The same property holds for the chain $F_1,..., F_k, F$ which connects A to B. The property holds for $k + 1$.

3. The graph models

The modelization of an ad hoc mobile network is not an easy task. Indeed the versatility of radio propagation in presence of obstacles, distance attenuation and mobility is the source of great difficulties. In passing, one should notice that mobility not only encompasses host mobility but also the mobility of the propagation medium. For example when a door is open in a building, then the distribution of links change. If a truck passes between two hosts it may switch down the link between them. In this perspective building a realistic model that is tractable by analysis is hopeless. Therefore we will focus on models dedicated to specific scenarios.

There are two kinds of scenarios: the indoor scenarios and the outdoor scenarios. For the indoor scenario we will use the random graph model. For the outdoor scenarios we will use the random unit graph model. The most realistic model lies somewhere between the random graph model and the random unit graph model.

3.1 The random graph model for indoor networks

In the following we consider a wireless indoor network made of N nodes. The links are distributed according to a random graph with N vertices and link probability is p. In other words, a link exists between two given nodes with probability p. The link's existence are independent from one pair of nodes to another. Figure 1 shows an example of a random graph with $(N, p) = (10, 0.7)$, the nodes having been drawn in concentric mode just for convenience.

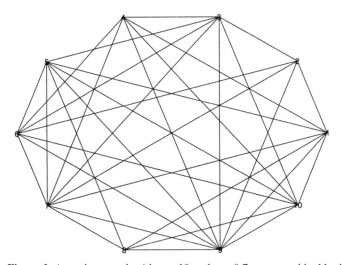

Figure 1. *A random graph with n = 10 and p = 0.7, generated by Maple*

The random graph model implicitly acknowledges the fact that in an indoor network, the main cause of link obstruction is the existence of a random obstacle (wall, furniture) between any pair of nodes. The fact that the links are independently distributed between node pairs assumes that these obstacles are independently distributed with respect to node position, which of course is never completely true. However the random graph model is the simplest satisfactory model of an indoor ratio network and provides excellent results as a starting point.

When the network is static, then the graph does not change during the time. It is clear that nodes do not frequently change position in the indoor model, but the propagation medium can vary. In this case the random graph may vary with time. One easy way to model the time variation is to assume random and independent link lifetime. For example, one can define μ as the link variation rate, i.e. the rate at which each link may come down or up. During an interval $[t, t + dt]$ a link can change its status with probability μdt, i.e. it takes status "up" or "down" with probability p, independently of its previous status. The effect of mobility will not be investigated in the present paper.

3.2 The random unit graph model for outdoor networks

To explain this kind of graph it suffices to refer to a very simple example. Let L be a non-negative number and let us define a two-dimensional square of size $L \times L$ unit lengths. Let us consider N nodes uniformly distributed in this square. The unit graph is the graph obtained by systematically linking pairs of nodes when their distance is smaller or equal to the unit length. This model of graph is well adapted to outdoor networks where the main cause of link failure is the attenuation of signal by distance. In this case the area where a link can be established with a given host is exactly the disk of radius equal to the radio range centered on the host. However the presence of an obstacle may give a more twisted shape to the reception area (that may not be singly connected).

Figures 2 and 3 respectively show the two steps of the build up of a random unit graph of dimension two. The first step is the uniform distribution of the points in the rectangle area. The second step is the link distribution between node pairs according to their distance.

The reception area may also change with the time, due to node mobility, obstacle mobility, noise or actual data traffic. In the present paper we will assume that the network is static.

Of course, the unit graph can be defined in other spaces than the plane. For example a unit graph can be defined on a 1D segment, modeling a mobile network made of cars on a road. It can be a cube in air, modeling a mobile network made of airplanes, for example.

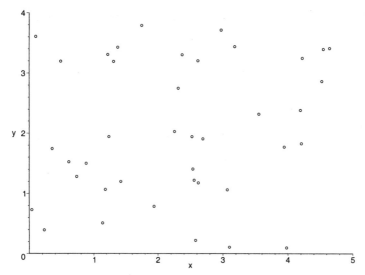

Figure 2. *Forty points uniformly distributed on a 5 × 4 rectangle*

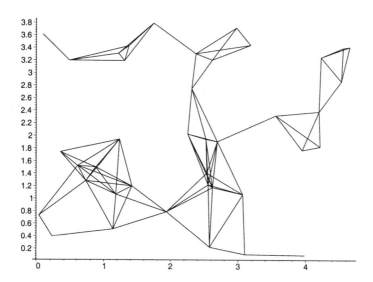

Figure 3. *The random unit graph derived from the forty points' locations of Figure 2*

4. Analysis of OLSR in the random graph model

4.1 Route lengths

Most pro-active protocols (like OLSR) have the advantage of delivering optimal routes (in terms of hop number) to data transfers. The analysis of optimal routes is very easy in random graph models since a random graph tends to be of diameter 2 when N tends to infinity with fixed p.

Theorem 3 *The optimal route between two random nodes in a random graph, when N tends to infinity,*

> *1. is of length 1 with probability p;*
> *2. or of length 2 with probability q = 1 – p.*

Proof: Point (1) is a simple consequence of the random graph model. For point (2) we consider two nodes, node A and node B, which are not at distance 1 (which occurs with probability $1 - p$). We assume *a contrario* that these nodes are not at distance 2, and we will prove that would occur with a probability p_3 which exponentially tends to zero when N increases. If the distance between A and B is greater than 2, then for each of the $N - 2$ remaining nodes in the network either

> 1. the link to A is down; or
> 2. the link to B is down.

For every remaining node the above event occurs with probability $1 - p^2$, therefore $p_3 = (1 - p^2)^{N-2}$, which proves the theorem.

4.2 Multipoint relay flooding

Theorem 4 *For all $\epsilon > 0$, the optimal MPR set size D_N of any arbitrary node is smaller than $(1 + \epsilon)\frac{\log N}{-\log q}$ with probability tending to 1 when N tends to infinity. (See Figure 4.)*

Proof: We assume that a given node A randomly selects k nodes in its neighborhood and we will fix the appropriate value of k which makes this random set a multipoint relay set. The probability that any given other point in the graph be not connected to this random set is $(1 - p)^k$. Therefore the probability that there exists a point in the graph which is not connected via a valid link to the random set is smaller than $N(1 - p)^k$. Taking $k = (1 + \epsilon)\frac{\log N}{-\log q}$ for $\epsilon > 0$ makes the probability tend to 0.

Notice that $D_N = 0 (\log N)$, which very favorably compares to the size of the whole host neighborhood (which is on average pN) and considerably reduces the topology broadcast.

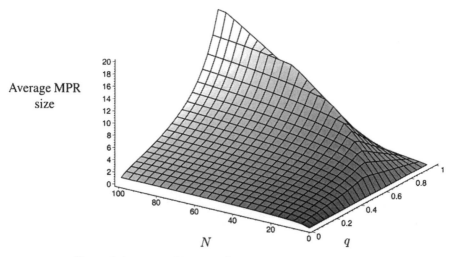

Average MPR size

Figure 4. Average multi-point relay set size in (p, N), q = 1 − p

Theorem 5 *The broadcast or flooding via multipoint relays takes on average a number R_N of retransmissions smaller than* $(1 + \epsilon) \dfrac{\log N}{-p \log q}$. *(See Figure 5.)*

Proof: First we notice that this average number favorably compares to the unrestricted flooding needed in plain links state routing algorithms which exactly need N retransmissions per flooding.

Let us consider a flooding initiated by an arbitrary node. We sort the retransmission of the original messages according to their chronological order. The 0-th retransmission corresponds to the source of the broadcast. We call m_k the size of the multipoint relay set of the k-th retransmitter. We assume that each of the m_k multipoint nodes of the kth transmitter are chosen randomly as in the proof of theorem 4. The probability that a given multipoint relay points of the k-th transmitter did not receive a copy of the broadcast packet from the k first retransmissions is $(1 - p)^k$. Therefore the average number of new hit multipoint relays which will have to actually retransmit the broadcast message after its kth retransmission is $(1 - p)^k m_k$. Consequently the average total number of retransmissions does not exceed $\sum_{k \geq 0} (1 - p)^k m_k$.

Using the upper bound $m_k \leq (1 + \epsilon) \dfrac{\log N}{-\log q}$, the average number of retransmissions is upper bounded by $(1 + \epsilon) \dfrac{\log N}{-\log q} \sum_{k \geq 0} (1 - p)^k = (1 + \epsilon) \dfrac{\log N}{-p \log q}$, which completes the proof of the theorem.

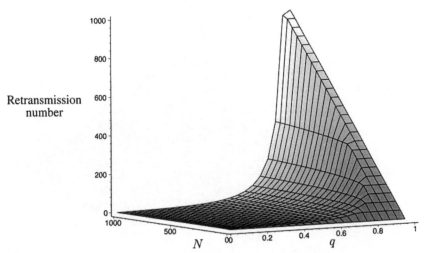

Figure 5. Average number of retransmissions in a multi-point relay flooding in (q, N)

Corollary 1 *The cost of OLSR control traffic for topology broadcast in the random graph model is $O(N(\log N)^2)$ compared to $O(N^3)$ with plain link state algorithm.*

4.2.1 Note

Notice that the neighbor sensing in $O(N^2)$ is now the dominant source of control traffic overhead.

5. Analysis of OLSR in the random unit graph

5.1 Analysis in 1D

A 1D unit graph can be made of N nodes uniformly distributed on a strip of land whose width is smaller than the radio range (set as unit length). We assume that the length of the land strip is L unit length.

Theorem 6 *The size of the MPR set D_N of a given host is 1 when the host is at one radio hop to one end of the strip, and 2 otherwise. (See Figure 6.)*

Proof: The proof is rather trivial. The heuristic finds the nodes which cover the 2-hop neighbors of the host. These nodes are the two nodes which are the farther from the neighborhood of the host (one on the left side, the other one on the right side). These two nodes cover the whole 2-hop neighborhood of the host and make the optimal MPR set. Notice that only one MPR suffices when the strip ends in the radio range of the host.

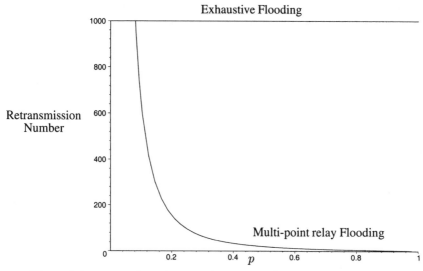

Figure 6. *Average number of retransmissions in multi-point relay flooding with N = 1000 and p variable*

Theorem 7 *The MPR flooding of a broadcast message originated by a random node takes $R_N = \lfloor L \rfloor$ retransmissions of the message when N tends to infinity and L is fixed. (See Figure 7.)*

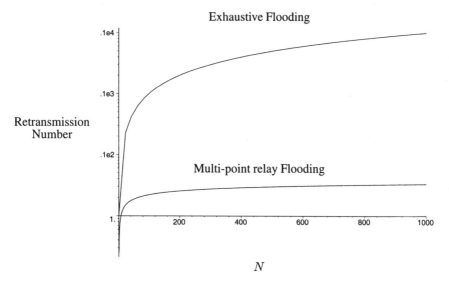

Figure 7. *Average number of retransmissions in multi-point relay flooding with N variable and p = 0.9, logarithmic scale*

Proof: The distance between the host and its MPR tends to be equal to one unit length when the density increases.

Notice this is assuming an error free retransmission. In case of error, the retransmission stops at the first MPR which does not receive the message correctly. In order to cope with this problem one may have to add redundancy in the MPR set which might be too small with regard to this problem.

Notice that these figures compare favorably with plain link state where $D_N = M = N/L$ and $R_N = N$.

5.2 Analysis in 2D

The analysis in 2D is more interesting because it gives less trivial results. We need the following elementary lemma about geometry. The proof is left to the reader.

Lemma 1 *Let consider two disks K_1 and K_2 of respective radius 1 and 2 centred on origin O. Let two points A and B be on the border of K_1 separated by an angle θ (measured from origin). Let $\mathcal{A}(\theta)$ be the area of the set of points of K_2 such that*

- *the points are not in K_1;*
- *the points are in the sector of origin O, limited by A and B;*
- *the points are at distance greater than 1 from both A and B.*

When $\theta \leq 2\pi/3$ then $\mathcal{A}(\theta) = \theta - \sin\theta$. Otherwise $\mathcal{A}(\theta) = \mathcal{A}(2\pi/3) + 3 \times (\theta - 2\pi/3)$.

Theorem 8 *When L is fixed and N increases, then the average size of the MPR set D_N tends to be smaller than $3\pi(N/(3L^2))^{1/3} = 3\pi(M/(3\pi))^{1/3}$.*

Notice that this figure compares favorably with plain link state where $D_N = M = N/L^2$.

Proof: We only give a sketch of the proof. We assume that L is not too small (greater than 4). Let consider an arbitrary host located on the square. Let K_1 be the disk of radius 1 centred on the host, and K_2 be the disk of radius 2 also centred on the host. We know that the one-hop neighborhood of the host is located in disk K_1, and the two-hop neighborhood is located in the set $K_2 - K_1$. To simplify, we assume that disk K_2 does not intercept the square border. The MPR selection heuristic naturally leads to select MPRs in the limit of the radio range of the host, i.e. the closer to the border of disk K_1. Indeed, this is where the neighbors cover most of the two-hop neighborhood of the host (i.e. $K_2 - K_1$). When the network density is high we can assume that the MPRs are actually on the border of K_1.

Let us consider k MPRs candidates identified by $B_1, \ldots, B_k$. We suppose that the B_i considered in increasing order are located clockwisely on the unit circle. Let θ_i be the angle made by the sector limited by B_i and B_{i+1}, with the boundary case θ_k is the angle made by the sector limited by B_k and B_1. We have $\theta_1 + \ldots + \theta_k = 2\pi$.

In order to make $\{B_1, \ldots, B_k\}$ a suitable MPR set, one needs the union of the disks $K(B_i)$ of radius 1 and center B_i contain the whole two-hop neighborhood of the host. This condition is fulfilled when the uncovered set $K_2 - K_1 - \cup_{i=1}^{i=k} K(B_k)$ does not contain any node of the network. If it is not the case therefore one has to add to $\{B_1,\ldots, B_k\}$ extra neighbor nodes that cover the nodes in $K_2 - K_1 - \cup_{i=1}^{i=k} K(B_k)$.

The area of the uncovered set is exactly $\mathcal{A}_k = \mathcal{A}(\theta_1) + \cdots + \mathcal{A}(\theta_k)$, therefore the average number of nodes in this area is $\mathcal{A}_k \times D$, where D is the density of the network ($D = N/L^2$). Therefore the average number of extra nodes is smaller than $\mathcal{A}_k D$. Therefore $k + \mathcal{A}_k D$ is an upperbound of D_N. Notice that for k given, quantity $\mathcal{A}_k$ is minimal when the θ_i's are all equal, namely when $\theta_i = 2\pi/k$ for all i. In this case $\mathcal{A}_k = k\mathcal{A}(2\pi/k)$ and

$$D_N \le k + k\mathcal{A}(2\pi/k)D. \tag{5}$$

Using $\mathcal{A}(\theta) \approx \theta^3/6$, we have $D_N \le k + (2\pi)^3 D/(6k^2)$. The right-hand side attains its minimum for $k = 2\pi(D/3)^{1/3}$ and it becomes $D_N \le 3\pi(D/3)^{1/3}$.

Figure 8 displays simulation results for dimension 2. The heuristic has been applied to the central node of a random 4×4 unit graph. The convergence in $M^{1/3}$ is clearly shown. Notice that in this very case the upper bound of D_N is at least greater by a factor 2 than actual values obtained by simulations. Figure 9 summarizes the results obtained for quantity D_N in the random graph model for dimension 1 and 2. The results for dimension 2 have been simulated.

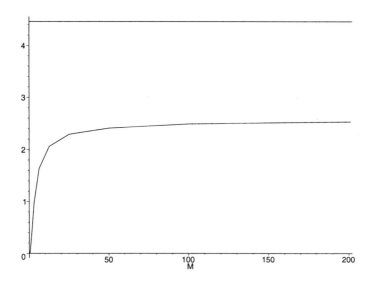

Figure 8. *Bottom: simulated quantity $D_N/M^{1/3}$ versus the number of neighbor M for the central position in a 4×4 random unit graph. Top: upper bound obtained in theorems*

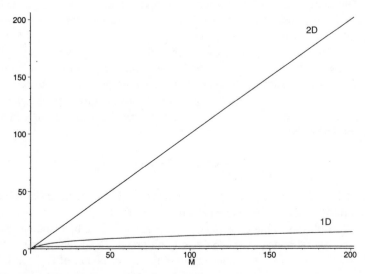

Figure 9. *Unit graph model from bottom to top, average number of MPR for 1D, 2D and*
full links state protocol versus the average number of neighbor nodes M

Theorem 9 *The MPR flooding of a broadcast message originated by a random*
node takes $R_N = O((NL^4)^{1/3})$ retransmissions of the message when N tends to
infinity and L is fixed.

Proof: There is no complete proof of this theorem. We can give a hint. We call
MPR hit when a retransmission is received by a MPR point for the first time. The
hit MPRs will have to retransmit the broadcast message. At each retransmission of
the broadcast packet there is an area size of order $O(1)$ which is added to the
already covered size. In the same time there are $O(M^{1/3})$ new additional MPR hits.
This new area contains $O(M)$ points. Therefore to cover the whole area there
would be a need of $O(N/M)$ retransmission with consequently $O(N/M \times (M)^{1/3})$
MPR hit. When the area is completely covered there is no possibility of new MPR
hit and flooding stops. The total number of retransmissions equals the number of
MPR hits which is $O(NM^{-2/3})$. The estimate $M = O(N/L^2)$ terminates the proof.

6. Conclusion and further works

We have presented a performance evaluation of OLSR mobile ad-hoc routing
protocols in the random graph model and in the random unit graph model. The
originality of the performance evaluation is that it is completely based on
analytical methods (generating function, asymptomatic expansion) and does not
rely on simulation software. The random graph model is realistic enough for
indoor or short range outdoor networks where link fading mainly comes from

random obstacles. The random unit graph model is realistic for long range outdoor networks where link fading mainly comes from distance attenuation. In this case the random graph model can be improved by letting the parameter p depend on distance x between the nodes. The analytical derivation of the performances of the routing protocol in the distance dependent random graph will be the subject of further work.

REFERENCES

[1] J. M. McQuillan, I. Richer, E. C. Rosen, "The New Routing Algorithm for the ARPANET", *IEEE Trans. Commun.* COM-28:711–719.

[2] D. B. Johnson, D.A. Maltz, "Dynamic Source Routing in Ad Hoc Wireless Networks", in *Mobile Computing*, Ch. 5, p. 153–181, Kluwer Academic Publishers, 1996.

[3] C. E. Perkins, E. M. Royer, "Ad Hoc On-Demand Distance Vector Routing", *IEEE Workshop on Mobile Computing Systems and Applications*, p. 90–100, 1999.

[4] M. S. Corson, V. Park, "Temporally Ordered Routing Algorithm", draft-ietf-manet-tora-spec-02.txt, 1999.

[5] M. Gerla, G. Pei, X. Hong, Tsu-Wei, "Fisheye State Routing Protocol (FSR) for Ad Hoc Networks", MANET draft, draft-ietf-manet-fsr-01.txt, 2000.

[6] P. Jacquet, P. Mühlethaler, A. Qayyum, A. Laouiti, L. Viennot, T. Clausen, "The Optimized Link State Routing Protocol", MANET draft, draft-ietf-manet-olsr-02.txt, 2000.

[7] B. Bellur, R. Ogier, F. Templin, "Topology Broadcast Based on Reverse-path Forwarding", draft-ietf-manet-tbrpf-01.txt, 2001.

[8] Phillipe Jacquet, Pascale Minet, Paul Mühlethaler, Nicolas Rivierre, "Increasing Reliability in Cable-free Radio LANs: Low Level Forwarding in HIPERLAN", in *Wireless Personal Communications*, vol. 4, no. 1, p. 51–63, 1997.

[9] L. Viennot, "Complexity Results on Election of Multipoint Relays in Wireless Networks", *INRIA RR-3898*, 1998.

[10] P. Jacquet, A. Laouiti, "Analysis of Mobile Ad Hoc Network Routing Protocols in Random Graphs", *INRIA RR-3835*, 1999.

[11] A. Qayyum, L. Viennot, A. Laouiti, "Multipoint Relaying: An Efficient Technique for Flooding in Mobile Wireless Networks", *INRIA RR-3898*, 2000.

[12] P. Jacquet, L. Viennot, "Overhead in Mobile Ad Hoc Network Protocols", *INRIA RR-3965*, 2000.

[13] A. Qayyum, *Analysis and Evaluation of Channel Access Schemes and Routing Protocols for Wireless Networks*, Thèse de l'Université Paris 11, 2000.

[14] P. Mühlethaler, A. Najid, "An Efficient Simulation Model for Wireless LANs Applied to the IEEE 802.11 Standard", *INRIA RR-4182*, 2001.

[15] P. Jacquet, A. Laouiti, P. Minet, L. Viennot, "Performance Analysis of OLSR Multipoint Relay Flooding in Two Ad Hoc Wireless Network Models", *INRIA RR-4260*, 2001.

Chapter 7

Providing Differentiated Services (*DiffServ*) in wireless ad hoc networks

J Antonio García-Macías, Franck Rousseau, Gilles Berger-Sabbatel, Leyla Toumi and Andrzej Duda
LSR-IMAG Laboratory, CNRS and Institut National Polytechnique de Grenoble, France

1. Introduction

In a typical ad hoc network, mobile hosts communicate over dynamically established routes using wireless links. Providing *Quality of Service (QoS)* in such networks is a difficult problem because of varying topology, dynamic routing, radio channel characteristics and complexity of mobility management. The problem is even much harder if wireless links are shared local area networks such as the basic DCF access method defined in IEEE 802.11: the access method (CSMA/CA) is designed to provide mobile hosts with a fair share of the channel capacity. Moreover, it has highly probabilistic behavior and incurs high overheads. In this paper, we explore how to provide QoS support over 802.11 wireless links in an ad hoc environment.

The current approach to providing quality of service in the global Internet is based on *Differentiated Services (DiffServ)* [Bla 98]. Its principle is to classify and mark up the traffic at the entrance of the backbone network so that it can be processed differently in backbone routers and obtain different performance for each assigned class. Performance of *DiffServ* relies on sufficient provisioning of network resources in the backbone. This model also assumes that resources in access networks (the networks between a host and the backbone) are over-provisioned as usually it is the case for current local area networks (LAN). However, if a mobile host is connected to a wireless LAN such as IEEE 802.11 or

Bluetooth, the radio channel becomes a critical part of the whole architecture and may severely affect the end-to-end performance. Although IEEE 802.11 provides a means for allocating a part of the radio channel bandwidth to some hosts (PCF – Point Coordination Function), we are interested in using the commonly available access method (DCF – Distributed Coordination Function) that is oriented towards fair sharing of the common communication channel. In this way, we use 802.11 as any other available link for transferring IP packets. In a previous paper [GAR 01], we proposed a QoS architecture in such an environment based on the *DiffServ* model. The architecture relies on *Access Routers* that manage wireless LAN cells and allocate resources to provide consistent IP level quality of service to mobile hosts.

In ad hoc networks, there is no central entity such as an *Access Router* that can manage resource allocations in a wireless LAN cell. However, we consider that all mobile hosts are in a close geographical neighborhood and at a given instant they form a group whose members cooperate to provide QoS differentiation. To extend our QoS architecture to this type of ad hoc networks, we propose to dynamically elect a *QoS Manager*, responsible for QoS allocations in a group of mobile hosts.

We start with the performance analysis of 802.11 wireless LAN links (Section 2), then we present *DiffServ* mechanisms (Section 3) and we give some details of experiments that show how we can achieve QoS differentiation over the 802.11b network (Section 4). Then, we present a dynamic scheme for managing QoS allocations in a one-hop ad hoc network (Section 5). Finally, we present related work (Section 6) and conclusions (Section 7).

2. Quality of service in IEEE 802.11b networks

A wireless LAN environment has specific characteristics that make it difficult to provide adequate quality of service. First of all, the 802.11 MAC layer raises the problem of the access overhead that increases with the number of active hosts. The DCF access method is based on the CSMA/CA principle in which a host wishing to transmit senses the channel, waits a period of time (DIFS – Distributed Inter Frame Space) then transmits if the medium is still free. If the packet is received correctly, the receiving host sends an ACK frame after another period of time (SIFS – Short Inter Frame Space). If the ACK frame is not received by the sending host, a collision is assumed to have occurred and the data packet is transmitted again after waiting another random amount of time.

If a single host transmits a data frame, the transmission time will be the following (we suppose 802.11b with the bit rate of 11 Mb/s [ANS 00] and we neglect propagation times; this analysis follows [CAL 98, Wei 97, BIA 00]):

$$T_{single} = t_{pr} + t_{tr} + SIFS + ACK + DIFS \qquad [1]$$

where t_{pr} is the preamble time (144 μs), t_{tr} is the frame transmission time (size/bit rate), SIFS = 10 μs, ACK is the ACK transmission time (210 μs), and DIFS = 30 μs. If we assume the frame size of 1500 bytes of data (data frame of total 1534 bytes), proportion r of the useful bandwidth in this case will be:

$$r = \frac{t_{tr}}{T_{single}} = \frac{1.11ms}{1.51ms} = 0.735 \qquad [2]$$

So, a single host sending over a 11 Mb/s radio channel will have the useful bandwidth of 8.08 Mb/s.

If there are multiple hosts attempting to access the channel, one host may sense a busy channel or collide with the transmission of another host. In such cases, the host executes the exponential backoff algorithms to wait a random interval distributed uniformly between $[0, CW - 1] \times slot$, $CW_{min} = 32$, $CW_{max} = 1024$, and $slot = 20$ μs (these parameters are for the direct sequence spread spectrum physical layer). Each time the host chooses a slot and happens to collide, it will double CW up to CW_{max}. So if there are m hosts, the efficiency degrades because of collisions.

$$T_{multiple}(m) = t_{pr} + t_{tr} + SIFS + ACK + DIFS + w_{cw}(m) \qquad [3]$$

where $w_{cw}(m)$ is the mean length of the contention window for m hosts. This means that the proportion of the maximum bandwidth will also depend on the number of hosts:

$$r(m) = t_{tr}/T_{multiple}(m) \qquad [4]$$

For example, if we assume 1500 bytes of data and one collision on average, i.e. $w_{cw}(m) = 0.31$ ms, the efficiency decreases to 0.61, and the useful bandwidth to 6.71 Mb/s. So, if we want to manage bandwidth allocations, we have to take into account the fact that the available bandwidth of the 802.11 link depends strongly on the number of active hosts and their traffic.

Fact 1. *To provide quality of service over the 802.11 link, the number of hosts allowed to use the channel should be limited.*

Another problem of 802.11 is related to the performance of the radio channel that is time and location dependent due to factors such as the distance between the source and the destination, signal interference and fading. Some wireless LANs make use of different modulation and error control techniques so that these factors manifest themselves as variation in bandwidth perceived at the network layer. However, the most popular 802.11 products do not provide such a support. Instead, they are able to degrade the bit rate when repeated frame drops are detected (e.g. WaveLAN can degrade from 11 Mb/s to 5.5, 2, or 1 Mb/s). However, as the channel access probability is equal for all hosts, hosts that send at low rates penalize hosts that use the high rate. Table 1 shows the measured performance of a 802.11b WLAN with two hosts that use different rates (the throughput is measured at the TCP layer). We can see that the low rate host penalizes the high rate host and both hosts obtain a small proportion of the nominal bandwidth.

Table 1. 802.11b performance, hosts of different rate

host rates	measured throughput
11 Mb/s, 11 Mb/s	5.0 Mb/s
11 Mb/s, 1 Mb/s	0.84 Mb/s

This means that if we want to provide a satisfactory QoS behavior, we have to restrict the usage of the 802.11 link to an area in which all hosts can send at the same high rate, e.g. 11 Mb/s.

Fact 2. *To provide quality of service over the 802.11 link, the geographical area in which mobile hosts communicate should be limited so that all hosts use the same high bit rate.*

The access method (CSMA/CA) of 802.11 is designed to provide mobile hosts with a fair share of the radio channel capacity. If we want to provide different performance behavior to traffic sources at mobile hosts, we need to constrain them in a configurable way so that sources of low priority benefit from different resource allocations than high priority ones. For example, we can use traffic shapers to constrain sources at mobile hosts and keep in this way the aggregated traffic lower than the available link capacity.

Fact 3. *To provide quality of service over the 802.11 link, traffic sources should be constrained by configuring traffic shapers in hosts to obtain desired QoS effects.*

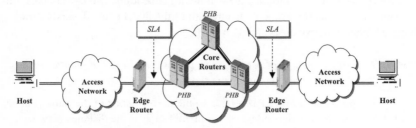

Figure 1. Architecture of differentiated services

In addition to that, QoS management should be reactive enough to adapt to varying conditions in an ad hoc network such as starting or terminating a traffic source, arrival or departure of a host in/from a group.

3. Differentiated Services

The architecture of *DiffServ* distinguishes two parts: the core network composed of one or several ISPs, packet forwarding done by core routers and the access network connecting end hosts to an edge router (cf. Figure 1). Performance agreements between different administrative domains (*SLA – Service Level Agreements*) allow us to statically reserve sufficient resources to support statistical performance guarantees of different traffic classes. Core routers forward packets according to different *BA (Behavior Aggregates) –* QoS classes that group flows of similar properties. Performance perceived by each class depends on the type of processing at core routers specified in a *PHB (Per Hop Behavior).* Edge routers perform classification of the incoming traffic and marking according to application types, source and destination addresses or ports or other criteria. Incoming traffic is checked against a TCA (*Traffic Conditioning Agreement*), a profile of the traffic defined in the SLA. Traffic exceeding a given TCA can be dropped, marked as out of profile, or marked with a lower priority class.

We use an implementation of the *DiffServ* edge and core router functions developed in a Next Generation Internet project [@IRS 01]. It is based on an IPv6 stack and has slightly different properties to those proposed by IETF. The differentiated services define three classes:

- EF *(Expedited Forwarding).* It provides flows with small delay and jitter as well as with low packet drop rate that is suitable for interactive real-time applications. To achieve such performance, EF packets have higher priority than other classes. EF flows are rate envelope multiplexed: waiting probability of EF packets is kept low by controlling the number of admitted flows based on their peak rate and by providing enough resources (link capacity).
- AF *(Assured Forwarding).* It defines a QoS class for elastic flows that do not have the strict requirements of EF flows, but need a minimum bandwidth. If the network is not congested, AF flows may obtain more bandwidth. To avoid confusion, we define only one AF class instead of four in *DiffServ* and two spatial priorities (drop probability thresholds) instead of three in *DiffServ.*
- BE *(Best Effort).* This class, which exists in the current Internet, does not provide any QoS guarantee.

The edge router functions are presented in Figure 2. Incoming packets are classified and marked with a DSCP *(Differentiated Services Codepoint).* TCA specifies rules for classification and metering. Shaping of the EF class is done by a FIFO queue with a small size. Some bursts can be tolerated; however packets arriving when the queue is full are dropped. Packets leave the queue according to a given peak rate. The TCA for the AF class contains a token bucket that defines the mean rate and burst tolerance. Traffic exceeding the rate is marked as out of profile and can be eliminated by core routers in case of congestion. The BE class is not controlled at all.

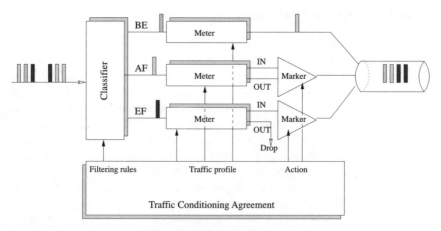

Figure 2. *Edge router functions*

The architecture of the core router is presented in Figure 3. It is composed of three queues for each class of the traffic. The EF class has a static priority higher than AF and BE. The AF and BE classes are scheduled according to a variant of WFQ *(Weighted Fair Queueing)*: WF2Q+ *(Worst-case Fair Weighted Fair Queueing)* [Zha 96]. The proportion of the bandwidth allocated to the AF and BE classes is configurable, for example 60% and 40%. We use a tail-drop policy for the EF and BE queues: a packet is dropped when the queue is full. Conformant and non-conformant packets of the AF class are subject to the PBS *(Partial Buffer Sharing)* policy: only conformant packets are accepted when the queue size is greater than a given threshold. In this way, all AF packets may benefit from available resources; however, in case of congestion only conformant packets will be allowed in the network. The output traffic is limited by a token bucket to fit the rate of the output link. Note that the EF class benefits from a fixed part of the available bandwidth and the AF and BE classes share the bandwidth not used by EF flows.

The implementation of *DiffServ* gives us a set of mechanisms to manage quality of service in a wireless LAN environment: classification and marking, packet scheduling and traffic shaping. All mobile hosts that want to benefit from the service differentiation need to implement the edge and core router functions.

4. Performance of *DiffServ* over 802.11b links

We have implemented the *DiffServ* mechanisms on FreeBSD 3.2 notebooks that use a shared 11 Mb/s 802.11b wireless link. The *DiffServ* mechanisms are implemented in the IPv6 stack so we were able to measure performance of service differentiation presented below.

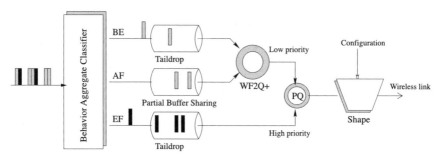

Figure 3. *Core router functions*

4.1 Measured performance

We have measured performance of service differentiation in the following experiment (cf. Figure 4). A mobile host has two traffic sources: a UDP source generating a priority EF traffic of rate 300 Kb/s with short 50 Bytes packets (a simple request-response test application) and a TCP source generating an elastic BE traffic (netperf tool for measuring useful bandwidth with 1 KB packets). In the first experiment, the QoS control mechanisms are inactive. Figure 5 presents the bandwidth obtained by the BE source measured at the application layer. The BE class is in competition with the EF class and gains most of the available bandwidth – we can see that its bandwidth stays around 5 Mb/s. We also show the round trip delay (RTT) of the EF class (cf. Figure 6). Until SeqNum = 100 the EF class is in competition with the greedy BE class. We can observe that the EF class is severely disturbed by the BE class, because both classes are scheduled according to the FIFO policy. At SeqNum = 100, the BE source stops sending, so that the RTT of the EF class becomes shorter, around 2.5 ms, and much more predictable.

The second experiment tests the isolation of both classes by means of the *DiffServ* control mechanisms. The output traffic shaper limits the bandwidth of the BE class to 2.4 Mb/s. Figure 7 shows the bandwidth obtained by the BE source, which is effectively maintained around 2.4 Mb/s. It can also be seen that the RTT of the EF class is much less disturbed by the BE class (cf. Figure 8). It is still greater than 2.5 ms, because of the competition with the BE class (the priority policy is not preemptive and an EF packet may wait an interval corresponding to the residual waiting time). As previously, the BE source stops sending in the middle of the observation (SeqNum = 130). These measures show that it is possible to isolate different QoS classes and obtain satisfactory performance.

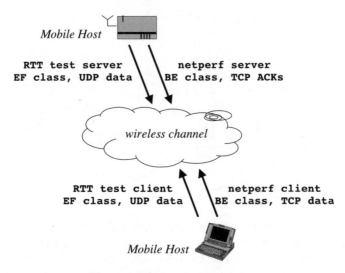

Mobile Host

RTT test server   **netperf server**
EF class, UDP data   **BE class, TCP ACKs**

wireless channel

RTT test client   **netperf client**
EF class, UDP data   **BE class, TCP data**

Mobile Host

Figure 4. *Experimentation set up*

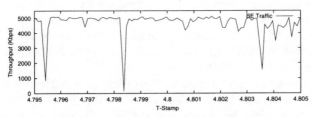

Figure 5. *No QoS control, bandwidth of BE class*

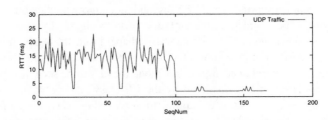

Figure 6. *No QoS control, RTT of EF class*

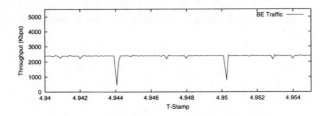

Figure 7. *QoS control, bandwidth of BE class*

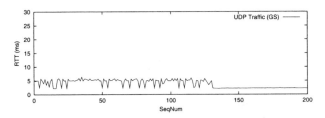

Figure 8. QoS control, RTT of EF class

5. Dynamic QoS allocation in ad hoc networks

Based on the results of the experiments, we propose a dynamic scheme for managing QoS allocations in ad hoc networks. At present, we limit the scope of our work to one-hop networks sharing the same 802.11 radio channel, because multi-hop networks rise more complex problems, especially due to the interaction of QoS management with routing. So, we assume that all mobile hosts in our ad hoc network can communicate directly with each other over a common radio channel and at a given instant they form a *QoS group* whose members cooperate to provide QoS differentiation (cf. Figure 9). This configuration is similar to one wireless LAN cell. However, there is no base station that centralizes all traffic.

As we have seen, providing *DiffServ* QoS requires some control of wireless links: geographical span, limited number of hosts, and constrained traffic sources. Before a mobile host may benefit from a higher priority performance class, or even before it is allowed to use the shared link, it has to obtain some bandwidth allocation from the group of hosts sharing the same wireless link. We assume that there is a *Qos Manager* in the group of hosts responsible for QoS allocations. It is dynamically elected based for example on the highest MAC address at a given instant. It is informed about the current required bandwidth of each mobile host, keeps track of the number of hosts in the group, and configures the parameters of their *DiffServ* mechanisms to obtain desired QoS behavior.

QoS allocation is done in the following way. A mobile host wanting to benefit from QoS provisioning must first enter a QoS group by broadcasting a request for a QoS allocation.

The required allocation of the bandwidth is interpreted according to the QoS class: for the AF class, it represents the minimum rate that can be obtained from WFQ scheduling; for the EF class it is the peak rate. The QoS manager interprets requests and satisfies them if possible by verifying the existing allocations. When the request is accepted, the QoS manger configures the *DiffServ* mechanisms in the requesting host and if needed, in other mobile hosts of the QoS group (for example if the new request requires the modification of the current QoS settings).

The QoS manager performs bandwidth allocation as follows: given available bit rate capacity C and $x_{i,class}$, traffic rate of class *EF, AF, BE* requested by source i,

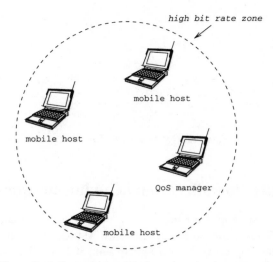

Figure 9. QoS group of mobile hosts in a one hop network

find proportions r_{EF}, r_{AF} and r_{BE} of the bandwidth to be allocated to each respective class:

$$x_{EF} = \sum x_{i,EF} \leq r_{EF}\ r(m)\ C$$
$$x_{AF} = \sum x_{i,AF} \leq r_{AF}\ r(m)\ C$$
$$x_{BE} = \sum x_{i,BE} \leq r_{BE}\ r(m)\ C \qquad [5]$$
$$r_{EF} + r_{AF} + r_{BE} = 1 - \delta$$

where δ accounts for over-provisioning of the allocation and $r(m)$ is the proportion of the effective bandwidth if m hosts are active.

To perform bandwidth allocation, we have measured the proportion of the useful bandwidth in function of the traffic load and the number of hosts in the 802.11b wireless LAN. Figure 10 presents the useful bandwidth for two and three competing hosts. Based on these statistics we can configure the *DiffServ* mechanisms of EF, AF and BE classes to limit their aggregated output rate to $r_{EF} r$ *(m) C,* $r_{AF} r$ *(m) C,* and $r_{EF} r(m)$ *C,* respectively.

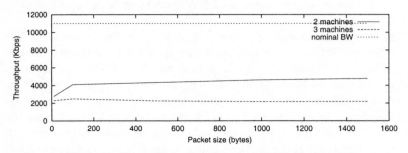

Figure 10. The useful bandwidth in the 802.11b wireless LAN

When the QoS manager decides that a new allocation requires configuration of the *DiffServ* mechanisms, it sends a configuration request to a mobile host. The mobile host configures the output rate of the EF and AF/BE classes and fixes the proportion between the AF and BE classes.

6. Related work

There is a large amount of work on QoS in ad-hoc networks [COR 98, COR 97], but many of them relate the problem of QoS to the problem of routing [COR 99, HSU 98, RAM 98, SIV 99]. Supporting IP QoS over wireless links is still an open problem addressed by the IETF community [MIT]. Recent surveys analyze different issues and identify research directions [Cha 99, Cha 00]. Our analysis follows their conclusions and applies them to the problem of providing *DiffServ* QoS in one hop ad hoc networks. Other work follows reservation-based QoS models such as RSVP [TAL 99]. Such an approach seems to be not suitable for highly dynamic ad hoc environments. Several authors have investigated providing service differentiation in 802.11 networks by extending or modifying the MAC layer [BAR 01, AAD 01]. However, these solutions cannot be applied to the networks that use current 802.11 products. Insignia defines an IP-based QoS framework for mobile ad hoc networks [LEE 98]. It is based on in-band signaling and soft-state resource management to support highly dynamic environments with time varying network topology, node connectivity, and end-to-end QoS. Its simple QoS model is based on providing mobile hosts with adaptive services: the allocation of a minimum bandwidth and the possibility to enhance to some maximum bandwidth.

7. Conclusions

This paper proposes to extend the differentiated services model to one hop ad hoc networks with wireless LAN links. We have analyzed the behavior of the 802.11b wireless LAN to state the principles of QoS management in a group of mobile hosts. To provide the required bandwidth to each performance class, we need to control several parameters of the wireless LAN link: a limited geographical span to ensure the same high bit rate for all communicating hosts, a constrained rate of traffic sources to limit the use of the radio channel in the function of the required QoS and a limited number of active hosts to keep the load sufficiently low. We have measured the performance of service differentiation between *DiffServ* classes and our first results show that it is possible to obtain sufficient isolation between these classes. Based on these experiments, we have proposed a dynamic scheme for managing QoS allocations in a group of mobile hosts. More work is needed to extend our approach to the case of multihop ad hoc networks.

Acknowledgements

This work has been supported by the French Ministry of Industry, National Network of Telecommunication Research (RNRT) via the @IRS project: *"Integrated Architecture for Networks and Services"*.

REFERENCES

[AAD 01] AAD I., CASTELLUCIA C., "Differentiation Mechanisms for IEEE 802.11", *INFOCOM*, 2001.

[ANS 00] ANSI/IEEE, "802.11: Wireless LAN Medium Access Control (MAC) and Physical Layer (PHY) Specifications", 2000.

[BAR 01] BARRY M., CAMPBELL A., VERES A., "Distributed Control Algorithms for Service Differentiation in WIreless Packet Networks", *INFOCOM*, 2001.

[BIA 00] BIANCHI G., "Performance Analysis of the IEEE 802.11 Distributed Coordination Function", *JSAC Wireless Series*, vol. 18, num. 3, 2000.

[Bla 98] BLAKE S., ET AL. "An Architecture for Differentiated Services", 1998, *Internet RFC 2475*.

[CAL 98] CALI F., CONTI M., GREGORI E., "IEEE 802.11 Wireless LAN: Capacity Analysis and Protocol Enhancement", *INFOCOM*, 1998.

[Cha 99] CHALMERS D., ET AL. "A Survey of Quality of Service in Mobile Computing Environments", *IEEE Online Communication Surveys*, vol. 2, num. 2, 1999.

[Cha 00] CHAN J., ET AL. "The Challenges of Provisioning Real-Time Services in Wireless Internet", *Telecommunications Journal of Australia*, vol. 50, num. 3, 2000.

[COR 97] CORSON M., "Issues in Supporting Quality of Service in Mobile Ad Hoc Networks", *IFIP Fifth International Workshop on Quality of Service (IWQOS '97)*, May 1997.

[COR 98] CORSON M., CAMPBELL A., "Toward Supporting Quality of Service in Mobile Ad Hoc Networks", *First IEEE Conference on Open Architecture and Network Programming (OPENARCH)*, April 1998.

[COR 99] CORSON M., MACKER J., "Mobile Ad Hoc Networking (MANET): Routing Protocol Performance Issues and Evaluation Considerations", January 1999, *Internet RFC 2501*.

[GAR 01] GARCÍA-MACÍAS J., ROUSSEAU F., BERGER-SABBATEL G., LEYLA T., DUDA A., "Quality of Service and Mobility for the Wireless Internet", *First ACM Wireless Mobile Internet Workshop*, Rome, Italy, 2001.

[HSU 98] HSU Y., TSAI T., LIN Y., "QoS Routing in Multihop Packet Radio Environment", *3rd IEEE Symp. on Computers and Communications (ISCC'98)*, 1998.

[@IRS 01] @IRS, "Integrated Architecture for Networks and Services", 2001, *RNRT Project*, http://www-rp.lip6.fr/airs/

[LEE 98] LEE S., CAMPBELL A., "INSIGNIA: In-Band Signaling Support for QoS in Mobile Ad Hoc Networks", *Mobile Multimedia Communications, MoMuC*, 1998.

[MIT] Mitzel D., "Overview of 2000 IAB Wireless Internetworking Workshop", *Internet RFC 3002*, 2000.

[RAM 98] Ramanathan R., Steenstrup M., "Hierarchically-organized, Multihop Mobile Wireless Networks for Quality-of-Service Support", *Mobile Networks and Applications*, vol. 3, num. 1, 1998.

[SIV 99] Sivakumar R., Sinha P., Bhargavan V., "CEDAR: A Core-Extraction Distributed Ad Hoc Routing Algorithm", *IEEE Journal on Selected Areas in Communications*, vol. 17, num. 8, 1999.

[TAL 99] Talukdar A., Badrinath B., Acharaya A., "MRSVP: A Reservation Protocol for Integrated Services Packet Networks with Mobile Hosts", *Wireless Networks*, vol. 5, num. 2, 1999.

[Wei 97] Weinmiller J., et al. "Performance Study of Access Control in Wireless LANs – IEEE 802.11 DFWMAC and ETSI RES 10 HIPERLAN", *Mobile Networks and Applications*, vol. 2, num. 1., 1997, p. 55–67.

[Zha 96] Zhang B., et al. "WF2Q: Worst-case Fair Weighted Fair Queueing", *INFOCOM 96*, 1996.

Index

Innovative Technology Series
Information Systems and Networks

Other titles in this series

Advances in UMTS Technology

Edited by J. C. Bic and E. Bonek
£58.00   1903996147   216 pages   April 2002

The Universal Mobile Telecommunication System (UMTS), the third generation mobile system, is now coming into use in Japan and Europe. The main benefits – spectrum efficient radio interfaces offering high capacity, large bandwidths, ability to interconnect with IP-based networks, and flexibility of mixed services with variable data – offer exciting prospects for the deployment of these networks.

This publication, written by academic researchers, manufacturers and operators, addresses several issues emphasising future evolution to improve the performance of the 3rd generation wireless mobile on to the 4th generation. Outlining as it does key topics in this area of enormous innovation and commercial investment, this material is certain to excite considerable interest in academia and the communications industry.

The content of this book is derived from *Annals of Telecommunications*, published by GET, Direction Scientifique, 46 rue Barrault, F 75634 Paris Cedex 13, France.

Java and Databases

Edited by A. Chaudhri
£35.00   1903996155   136 pages   April 2002

Many modern data applications such as geographical information systems, search engines and computer aided design systems depend on having adequate storage management control. The tools required for this are called persistent storage managers. This book describes the use of the programming language Java in these and other applications.

This publication is based on material presented at a workshop entitled 'Java and Databases: Persistence Options' held in Denver, Colorado in November 1999. The contributions represent the experience acquired by academics, users and practitioners in managing persistent Java objects in their organisations.

For information about other engineering and science titles published by Hermes Penton Science, go to **www.hermespenton.com**

Quantitative Approaches in Object-oriented Software Engineering

Edited by F. Brito e Abreu, G. Poels, H. Sahraoui, H. Zuse

£35.00   1903996279   136 pages   April 2002

Software internal attributes have been extensively used to help software managers, customers and users characterise, assess and improve the quality of software products. Software measures have been adopted to increase understanding of how software internal attributes affect overall software quality, and estimation models based on software measures have been used successfully to perform risk analysis and to assess software maintainability, reusability and reliability. The object-oriented approach presents an advance in technology, providing more powerful design mechanisms and new technologies including OO frameworks, analysis/design patterns, architectures and components. All have been proposed to improve software engineering productivity and software quality.

The key topics in this publication cover metrics collection, quality assessment, metrics validation and process management. The contributors are from leading research establishments in Europe, South America and Canada.

Turbo Codes: Error-correcting Codes of Widening Application

Edited by M. Jézéquel and R. Pyndiah

£50.00   1903996260   206 pages   May 2002

The last ten years have seen the appearance of a new type of correction code – the *turbo code*. This represents a significant development in the field of error-correcting codes.

The decoding principle is to be found in an iterative exchange of information (*extrinsic information*) between elementary decoders. The turbo concept is now applied to block codes as well as other parts of a digital transmission system, such as detection, demodulation and equalisation.

Providing an excellent compromise between complexity and performance, turbo codes have now become a reference in the field, and their range of application is increasing rapidly to mobile communications, interactive television, as well as wireless networks and local radio loops. Future applications could include cable transmission, short distance communication or data storage.

This publication includes contributions from an internationally-based group of authors, from France, Sweden, Australia, USA, Italy, Germany and Norway.

The content of this book is derived from *Annals of Telecommunications*, published by GET, Direction Scientifique, 46 rue Barrault, F 75634 Paris Cedex 13, France.

For information about other engineering and science titles published by Hermes Penton Science, go to **www.hermespenton.com**

Millimeter Waves in Communication Systems

Edited by M. Ney
£50.00   1903996171   180 pages   May 2002

The topics covered in this publication provide a summary of major activities in the development of components, devices and systems in the millimetre-wave range. It shows that solutions have been found for technological processes and design tools needed in the creation of new components. Such developments come in the wake of the demands arising from frequency allocations in this range. The other numerous new applications include satellite communication and local area networks that are able to cope with the ever-increasing demand for faster systems in the telecommunications area.

The content of this book is derived from *Annals of Telecommunications*, published by GET, Direction Scientifique, 46 rue Barrault, F 75634 Paris Cedex 13, France.

Intelligent Agents for Telecommunication Environments

Edited by D. Gaïti and O. Martikainen
£35.00   1903996295   110 pages   June 2002

Telecommunication systems become more dynamic and complex with the introduction of new services, mobility and active networks. The use of artificial intelligence and intelligent agents, integrated reasoning, learning, co-operating and mobility capabilities to provide predictive control are among possible ways forward. There is a need to investigate performance, flow and congestion control, intelligent control environment, security service creation and deployment and mobility of users, terminals and services. New approaches include the introduction of intelligence in nodes and terminal equipment in order to manage and control the protocols, and the introduction of intelligence mobility in the global network. These tools aim to provide the quality of service and adapt the existing infrastructure to be able to handle the new functions and achieve the necessary co-operation between nodes. This book's contributors, who come from research establishments all over the world, address these problems and provide ways forward in this fast-developing area of intelligence in networks.

For information about other engineering and science titles published by Hermes Penton Science, go to **www.hermespenton.com**

Video Data

Edited by M-S Hacid and S. Hassas
£35.00   1903996228   128 pages   July 2002

With recent progress in computer technology and reduction in processing costs it is possible to store huge amounts of video data needed in today's communication applications. To obtain efficient use of such data efficient storage, querying and navigation of this data is needed. To meet the increasing demands of the new developments, new management techniques and tools need to be developed, and this publication addresses the application of the many research disciplines involved.

Multimedia Management

Edited by J. Neuman de Souza and N. Agoulmine
£40.00   1903996236   140 pages   July 2002

With the arrival of multimedia services via the network, the study of multimedia transmission over various network technologies has been the focus of interest for research teams all over the world.

The previously antagonistic QoS and IP-based network technologies are now part of an integrated approach, which are expected to lead to new intelligent approaches to traffic and congestion control, and to provide the end user with quality service customised multimedia communications. This publication emanates from papers presented at a Multimedia Management conference held in Paris in May 2000.

For information about other engineering and science titles published by Hermes Penton Science, go to **www.hermespenton.com**

Applications and Services in Wireless Networks

Edited by H. Afifi and D. Zeghlache
£58.00   1903996309   260 pages   July 2002

Emerging wireless technologies for both public and private use have led to the creation of new applications. These include the adaptation of current network management procedures and protocols and the introduction of unified open service architectures. Aspects such as accounting for multiple media access and QoS (Quality of Service) profiling must also be introduced to enable multimedia service offers, service management and service control over the wireless Internet. Security and content production are needed to foster the development of new services while adaptable applications for variable bandwidth and variable costs will open new possibilities for ubiquitous communications. In this book the contributors, drawn from a broad international field, address these prospects from the most recent perspectives.

Mobile Agents for Telecommunication Applications

Edited by E. Horlait
£35.00   1903996287   110 pages   July 2002

Mobile agents are concerned with self-contained and identifiable computer programs that can move within a network and can act on behalf of the user and another entity. Most current research work on the mobile agent paradigm has two general goals: the reduction of network traffic and asynchronous interaction, the object being to reduce information overload and to efficiently use network resources. The international contributors to this book provide an overview of how the mobile code can be used in networking with the aim of developing further intelligent information retrieval, network and mobility management, and network services.

For information about other engineering and science titles published by Hermes Penton Science, go to **www.hermespenton.com**